IMAGES
of America

Covina Valley Citrus Industry

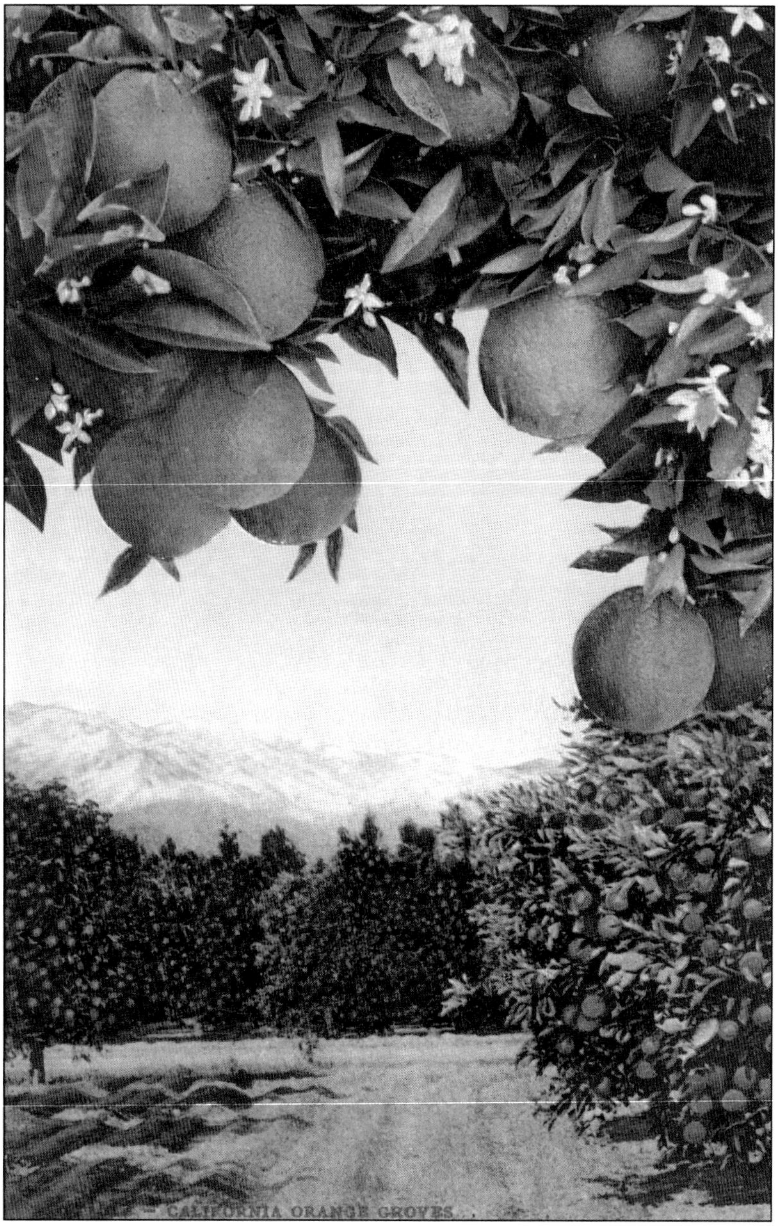

Snow-covered Mount San Antonio viewed through a grove of blooming Washington Navel oranges symbolizes the Covina Valley citrus industry. Gaspar de Portola, the first Spanish explorer to enter the San Gabriel Valley, named the mountain for Saint Anthony of Padua, his patron saint. The bountiful crop of golden fruit pictured offered the hope of prosperity and health to thousands who followed their dreams to California. (Courtesy Elliott-Milliken collection.)

ON THE COVER: It took 10 years for orange trees to reach this height. The pickers who came from Azusa, Glendora, Irwindale, and San Dimas used clippers to cut the oranges, which they placed in a canvas bags carried over their shoulders. When the bag was full, they emptied it into a field box. They initialed these boxes so the field boss could keep track of how many fruits they picked. (Photograph by Clarence W. Tucker; courtesy Covina Valley Historical Society.)

IMAGES
of America

COVINA VALLEY CITRUS INDUSTRY

Barbara Ann Hall, Ph.D.

ARCADIA
PUBLISHING

ISBN 978-0-7385-7442-4

Published by Arcadia Publishing
Charleston, South Carolina

Printed in the United States of America

Library of Congress Control Number: 2010939090

For all general information, please contact Arcadia Publishing:
Telephone 843-853-2070
Fax 843-853-0044
E-mail sales@arcadiapublishing.com
For customer service and orders:
Toll-Free 1-888-313-2665

Visit us on the Internet at www.arcadiapublishing.com

To my grandchildren:
Ashleigh Elizabeth and John Russell Horstmann;
Matthew Thomas Hubbard; and Michael Patrick, Jacqueline Elizabeth,
Gavin Christopher, and Garrett Winfield McGrade

Contents

Acknowledgments		6
Introduction		7
1.	Origins of Southern California Citrus	9
2.	Citrus Communities and the Water Wars	15
3.	First Citrus Cooperative and Pioneer Nurserymen	23
4.	New Citrus Cooperatives and Independent Packers	33
5.	Maintaining Commercial Citrus Groves	65
6.	Golden Years of Valley Citrus	75
7.	New Citrus Products	111
8.	Citrus Era Ends	119

ACKNOWLEDGMENTS

The author wishes to thank the following people for the generous support and expert assistance they have given to the preparation of this book: my Arcadia Publishing editor, Debbie Seracini; for image preparation, Marty Getz (Powell Camera); for manuscript preparation and proofing, Ashleigh Horstmann; and for Internet and library research, Wanda and Bob Updegraff. The author would also like to thank Karen Shomber, Dale Martin, Joseph Jones (Azusa Historical Society), Denise Bresee (curator, Glendora Historical Society), Jeffrie Ann Hubbard (archivist, Glendora Historical Society), Rosemarie Lippman (curator, Covina Valley Historical Society), Ida and Joe Fracasse (publishers, the *Glendora Magazine*), Win Patterson, Betty and Mark Thiel (Powell Camera), Mary Daniels Booth, and Leonard George Daniels III. A very special thank-you goes to Robert Ihsen, president of the Covina Valley Historical Society, for gathering and cataloging the societies' citrus collection.

Unless otherwise noted, photographs, postcards, and orange crate labels were published with permission from the following collections: the Azusa Historical Society, the Covina Valley Historical Society, the Glendora Historical Society, Wesley Berry, the Leonard George Daniels III family, Alice and Jerome Fahey, Pat Reynolds Milliken, Jack Milliken, and Claire and Bill Stone.

INTRODUCTION

Covina Valley Citrus Industry tells the story of how 25,000 acres of dry, rocky land covered with cactus and sagebrush were turned into a garden paradise by settlers who came from all over the United States. These men and women developed a successful commercial citrus industry that flourished for 70 years. According to an issue of the *Covina Argus* dated Saturday October 5, 1907, "Orange trees brought dollars for the sustenance of the people." The challenges were great. The first and most important obstacle was the need to develop an irrigation system that could function during periodic droughts. Dams to control floodwaters that periodically washed out groves were also a critical need. The industry required a transportation system that could efficiently reach distant markets with perishable product, and a national advertising campaign was needed to broaden the market. To function efficiently, the citrus packinghouses needed infrastructure such as telephone, gas, and electrical services. The valley communities were not large enough to develop these utilities on their own, so they developed services jointly. The citrus growers also worked together to create a cemetery and the first unified high school district in California.

Because the valley population was too small for a doctor to establish a practice, citrus growers formed a prepaid medical plan that hired a young doctor from Los Angeles to live in the valley. A home was provided for him, and he was paid a fixed fee per year for every family that signed up for the service. These examples of community cooperation to achieve economic and social goals helped the citrus industry grow.

The valley, which was originally called the Azusa Valley, later the Covina Valley, and finally the East San Gabriel Valley, is located south of the San Gabriel Mountains, north of the San Jose Hills and east of the San Gabriel River. During the California mission rancho period, it was part of the land owned by three families: the Daltons, the Rowlands, and the Workmans. In 1842, John Rowland purchased the Rancho La Puente from the Mexican government and shared it with partner William Workman. The part of the Rancho La Puente that is in the valley includes large areas of unincorporated land as well as the cities of Covina and West Covina.

In 1843, Henry Dalton, a British businessman from Peru, bought part of Rancho San Jose, the San Jose Addition, and the Rancho El Susa (renamed the Rancho Azusa del Dalton) from Ignacio Palomares, Luis Arenas, and Ricardo Vejar. The San Jose Addition includes the cities of Azusa and Irwindale. Dalton planted the first citrus in the valley. For his orchard, he bought seedling orange and lemon trees from the Mission San Gabriel Arcángel. He used a ditch built by Luis Arenas to irrigate his orchard with water from San Gabriel River.

Settlers came by horseback and wagon train to settle in the Azusa Valley. They bought and leased land near water sources from Henry Dalton, and he allowed them to use water from his ditch. With their help, Dalton doubled the capacity of the ditch and extended it to Broadway (presently Gladstone) Street. In 1858, an erroneous US government survey of Dalton's land changed his relationship with the settlers forever. The survey, conducted by Henry Hancock, took 18,500 acres of Dalton's best farmland and declared it government land. Many settlers who were leasing from Dalton filed homestead claims on the land they were farming. The results of the bitter, prolonged legal battle between Dalton and the settlers is best described by Sheldon G. Jackson, author of *A British Ranchero in Old California*: "The injustice of Henry Dalton's loss, the plight of the Azusa, and the poverty of the settlers stand as a mute indictment of United States Government land policies in California. In this case, everyone was a loser. For if the community had been permitted to develop naturally with clear titles to the land, the area would no doubt have been in a more prosperous condition by 1880. Instead, prolonged litigation brought financial adversity to both sides of the controversy, and the government lost an undetermined amount of potential taxes."

The controversy between Dalton and the settlers slowed the progress of agriculture in the valley. Small groves of seedling citrus trees were planted around Hanes's general store on Azusa Avenue and Eugene Griswold's store and meeting hall at present-day Cypress Street and Citrus Avenue. There was a market for citrus at these stores; miners heading to the San Gabriel Mountains in search of gold would pay a dollar a piece for lemons. A small settlement called Citrus grew up around Griswold's store. Mail for the valley settlers was delivered there. Today, envelopes with a Citrus postmark are very rare collectors' items.

The navel oranges exhibited in 1879 at the Riverside Citrus Fair gave a great impetus to the industry. In 1881, J.B. Beardslee planted the first navel orange trees on North Citrus Avenue. Switching from seedling oranges to grafted ones was the next important step for citrus industry development. The winter-blooming Riverside Navel orange (later called the Washington Navel) and the summer-blooming Valencia orange became industry staples. In 1887, when Glendora founder Charles Dexter Whitcomb brought the Atchison, Topeka, and Santa Fe Railroad to Glendora and Azusa, it became possible to ship citrus to lucrative markets in the Midwest and Eastern United States. The invention of the ventilated freight car in 1887 and the ice bunker car in 1889 greatly reduced losses due to delay in transit. At first, independent shipping agents bought the ranchers' crops and handled picking, packing, and marketing. Citrus distribution to the wholesale market was disorganized; some cities received too much product, while others received too little. The depression of the 1890s reduced the income taken by shipping agents, causing them to change their buying arrangements with the ranchers. They took crops on consignment and only paid the ranchers for the portion of the crop that was sold. The shipping agents also began charging the ranchers for picking, packing, and shipping services. Many ranchers were almost ruined.

In August 1893, Covina Valley citrus ranchers met at Slausons Hall in Azusa and formed the Azusa Covina Glendora (ACG) Citrus Association. It was the first stockholder-owned citrus cooperative. Each stockholder was requested to buy one share of stock for each acre of bearing trees. The association leased land from the railroad and built two small packinghouses, one in Azusa and one in Glendora.

Covina Valley ranchers became part of a growing movement throughout the California citrus industry to join in nonprofit cooperatives to market their fruit. T.H.B. Chamberlin of Riverside championed the movement. He and 11 friends and neighbors created the Pachappa Orange Growers Association near Riverside. In 1892, they incorporated without capital stock under the Benevolent Corporation of California law. At first, they contracted with a nearby packinghouse to process their fruit, and in 1895, they secured a packinghouse and began packing their own fruit.

On April 4, 1893, a group of almost 100 orange growers, including P.J. Dreher, president of the Claremont Fruit Growers Association, met in Los Angeles to discuss Chamberlin's plan for a Southern California cooperative association. On August 29, 1893, the Southern California Fruit Exchange was born.

One

Origins of Southern California Citrus

Citrus, a genus of the family Rutaceae native to India and other warm parts of Asia, includes oranges, citrons, lemons, limes, bergamots, shaddocks, and pomelos. Citrus is mentioned in writings that predate the Christian era. The first monograph about citrus cultivation was written in 1178 by Han Yen-Chih. He describes 27 varieties of oranges grown around Wen-Chou, China. Yen-Chih also wrote about grafting, transplanting, fertilizing, irrigating, fungus diseases, and boring insects. He described proper fruit picking and handling to prevent decay. Sweet oranges appeared in Europe early in the 15th century and were grown mostly in southern Europe. Spanish explorers recorded citrus growing in the West Indies and Brazil before 1600. Franciscan fathers who built the 21 California missions planted citrus in California between 1769 and 1823.

The Mission San Gabriel Arcángel had six acres of 400 seedling orange and lemon trees. Citrus cultivation declined at the missions after they were secularized in 1834. In 1841, William Wolfskill developed the first commercial citrus grove; he planted orange seeds from the Mission San Gabriel Arcángel on Central Avenue and Fifth Street in Los Angeles. This orchard was later increased to 70 acres. In the 1870s, budded varieties of citrus were introduced; they would replace seedling trees. The winter Washington Navel and the summer Valencia oranges became the chief commercial citrus. The Washington Navel orange originated in Goa, a Portuguese possession on the west coast of India. The Valencia orange originated in the Azores, a group of volcanic islands located in the Atlantic Ocean. The Lisbon lemon was imported from Australia, and the Eureka lemon was developed in Los Angeles, grown from a Sicilian lemon seed that produced higher quality fruit than its parent and sister trees.

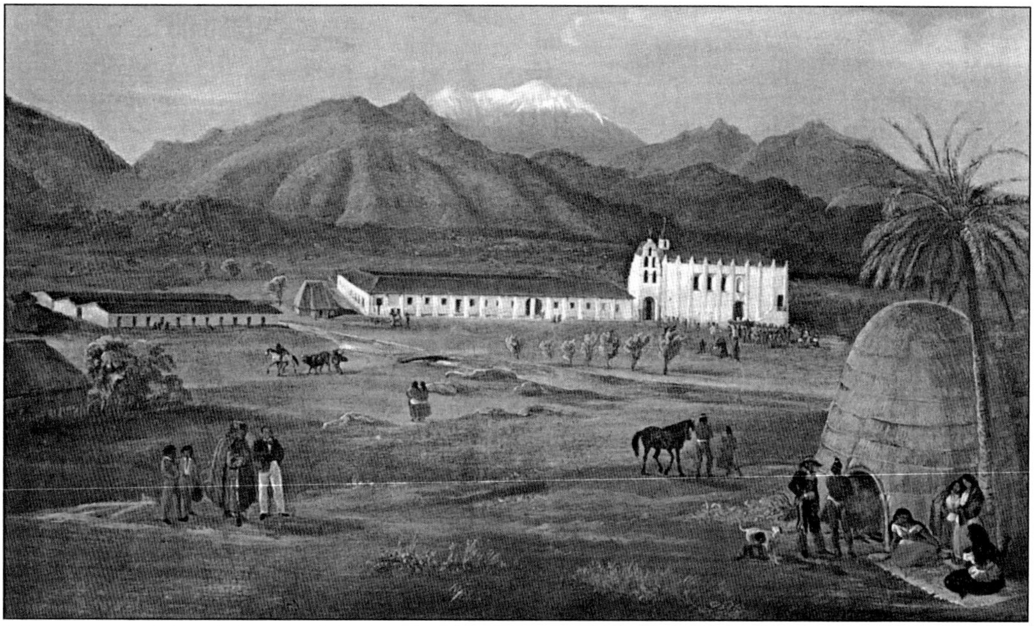

Mission San Gabriel Arcángel had 400 seedling citrus trees to supply fruit for Spanish galleons carrying Mexican gold back to Spain. It was important for sailors to have fresh citrus to prevent scurvy on long sea voyages. The success of San Gabriel's orchards proved that Southern California soil and climate were suitable for citrus culture. In 1828, artist Ferdinand Deppe visited the mission. He completed this painting, entitled *San Gabriel Mission*, from memory in 1832.

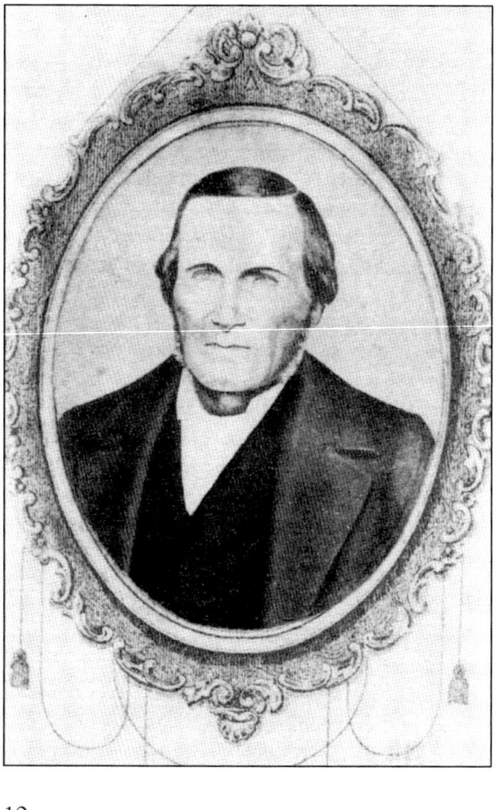

Famous frontier trapper William Wolfskill settled in Pueblo de Los Angeles and started cultivating Southern California's commercial citrus industry. By 1857, he had 2,000 trees, the largest orange orchard in the United States. By 1862, there were 25,000 citrus trees in California, and two-thirds of them were in Wolfskill's orchards. He pioneered local fruit sales and later shipped it to San Francisco using sailing vessels or steamers.

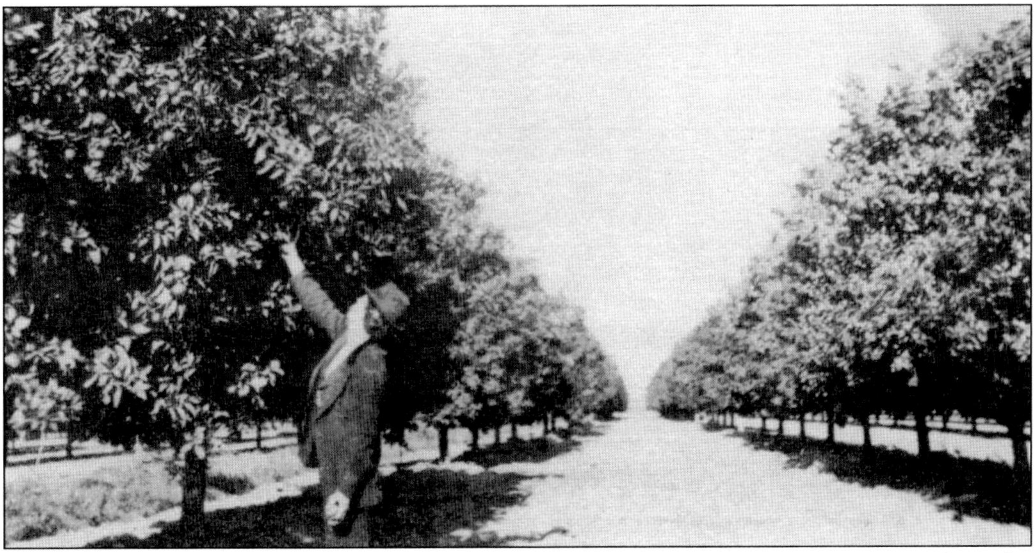

Wolfskill irrigated his orchards through a system of *zantas* (canals) built to carry water from the river. In 1880, Leland Stanford arranged for Pres. Rutherford B. Hayes to visit Wolfskill's orchards. An arch of evergreens with the monogram *H* was placed over the entrance, and the driveway was lined with American flags. The president reached from his carriage and picked an orange. In 1886, Wolfskill introduced the summer Valencia orange to his orchards.

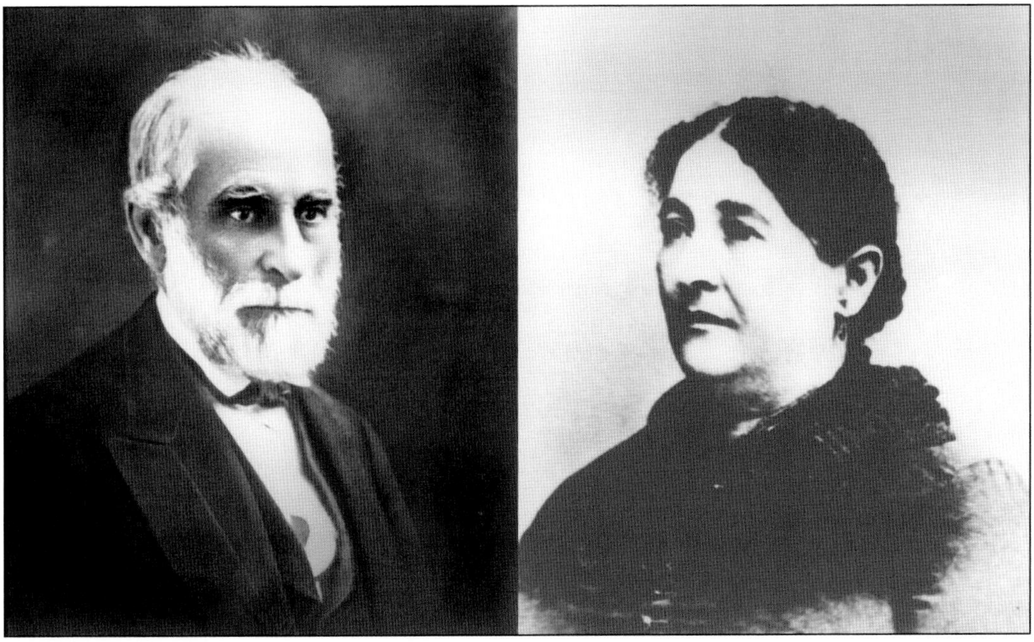

In 1856, Henry Dalton planted on his Rancho Azusa groves seedling orange and lemon trees from the San Gabriel Mission Arcángel. Dalton, a successful British businessman from Peru, acquired 45,280 acres of what was then called the Azusa Valley. In 1847, he married Maria Guadalupe Zamorano, whose father had served as governor of California. The gold rush in the San Gabriel Mountains brought many customers for his lemons.

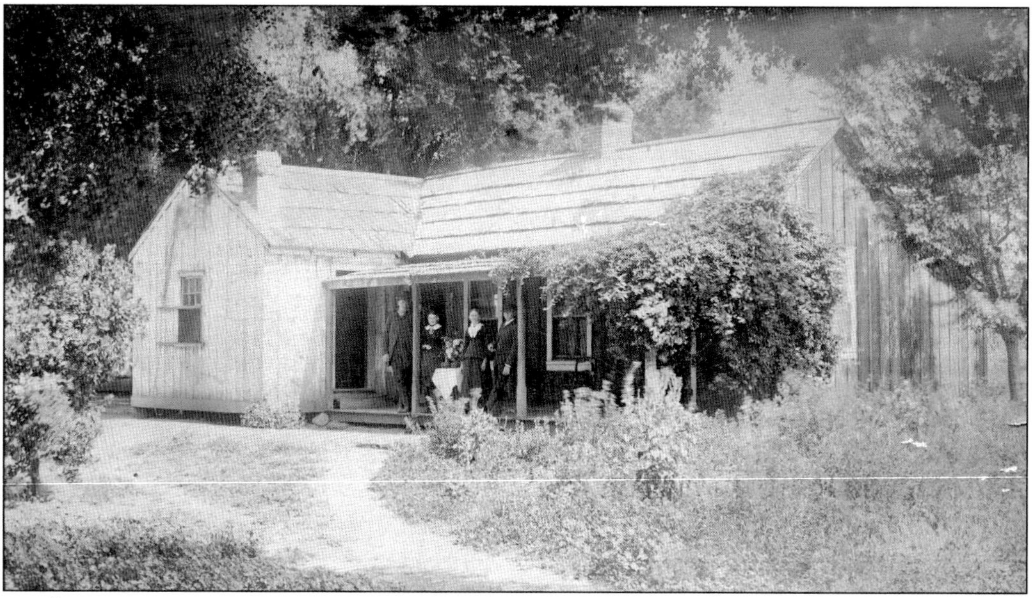

Seedling citrus trees were planted around Eugene Griswold's store and meeting hall at present-day Citrus Avenue and Cypress Street. The tiny settlement that grew up around the store was called Citrus, and residents received their mail at Griswold's. In 1868, the Charles Dougherty family bought a small ranch near Citrus from Henry Dalton and built this house. The Dougherty family brought their seven children and three orphaned cousins from Texas to California by wagon train.

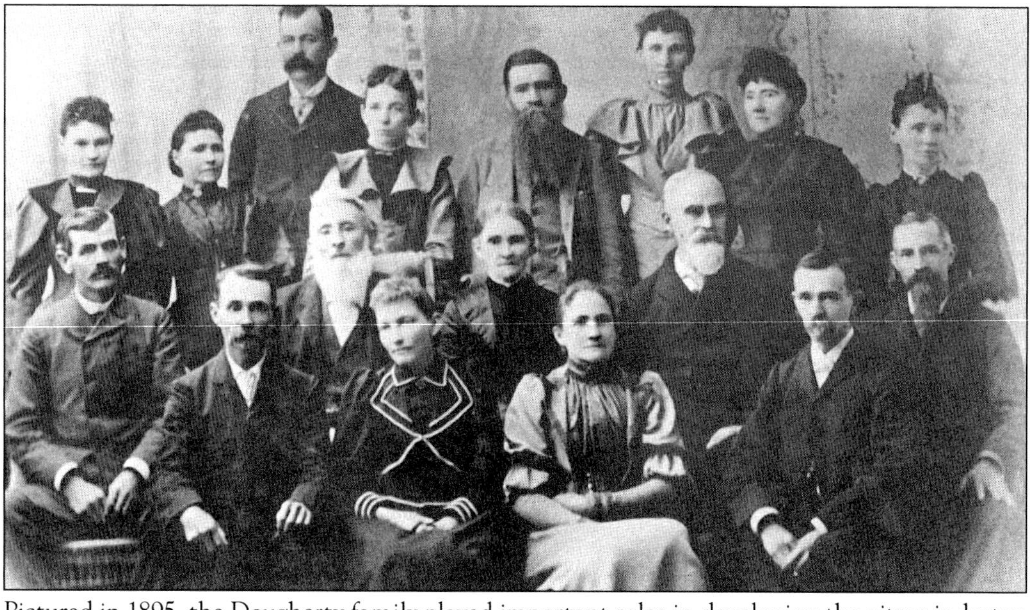

Pictured in 1895, the Dougherty family played important roles in developing the citrus industry. From left to right are (first row) Elgin Knott, William Dougherty, Mamie Reeves Dougherty, Callie Barlin Dougherty, James Louis Dougherty, and John Price; (second row) James Preston, Rosamond Dougherty, and John King; (third row) Jennie Dougherty Knott, Mollie Dougherty Preston, James Elliott, Mary Octavia and John Lindsay Dougherty, Carrie Griswold Elliott, Martha Dougherty King, and Nancy Dougherty Price.

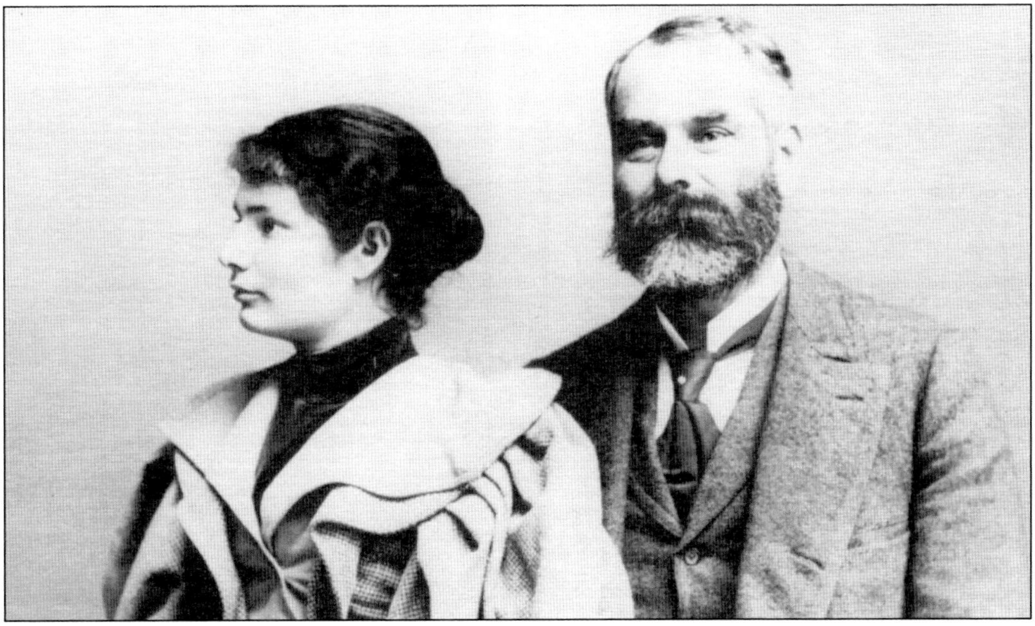

Pictured here with daughter Elena, Joseph Wolfskill worked with his father, William, producing citrus and developing new markets. In 1877, the Wolfskills sent the first shipment of oranges from Los Angeles to St. Louis. Each orange was wrapped in paper, and the boxes were packed in ice that had to be replaced 11 times before the trip ended.

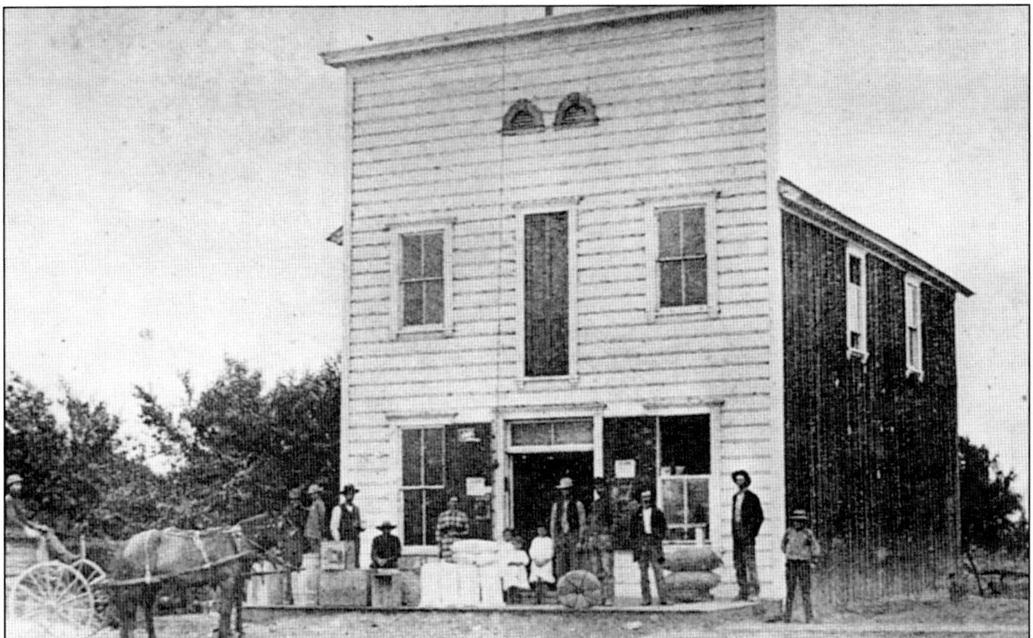

A citrus grove surrounds the first general merchandise store in the upper Azusa Valley; it was built by Phillip and Fred Hanes in 1886. In 1892, the store was moved to Azusa Avenue and Gladstone. Pictured from left to right are the following: Jesse Danks in the wagon; two unidentified people; Fred Hanes in shirtsleeves; unidentified; Frances Hanes (in plaid); sisters Barbara, Clara, and May Hanes; Phillip Hanes (on the left behind the roll of wire); and four unidentified.

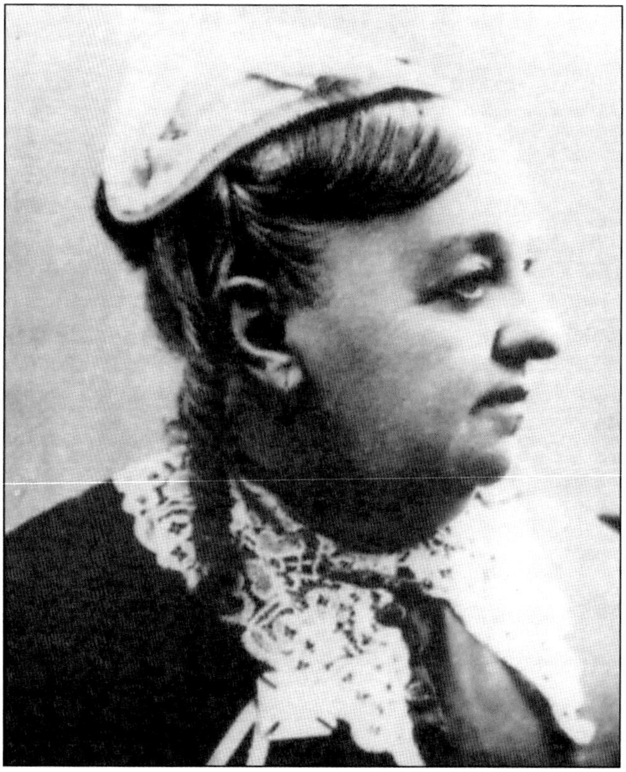

Eliza Tibbets of Riverside, California, received two special orange trees in the early 1870s. When William Saunders, the superintendent of the United States Department of Agriculture Gardens, learned of a superior orange variety grown in Bahia, Brazil, he ordered several trees. The first shipment from Bahia arrived dead, but in 1870, twelve small trees arrived in good condition. They were planted and propagated by budding on small orange stocks.

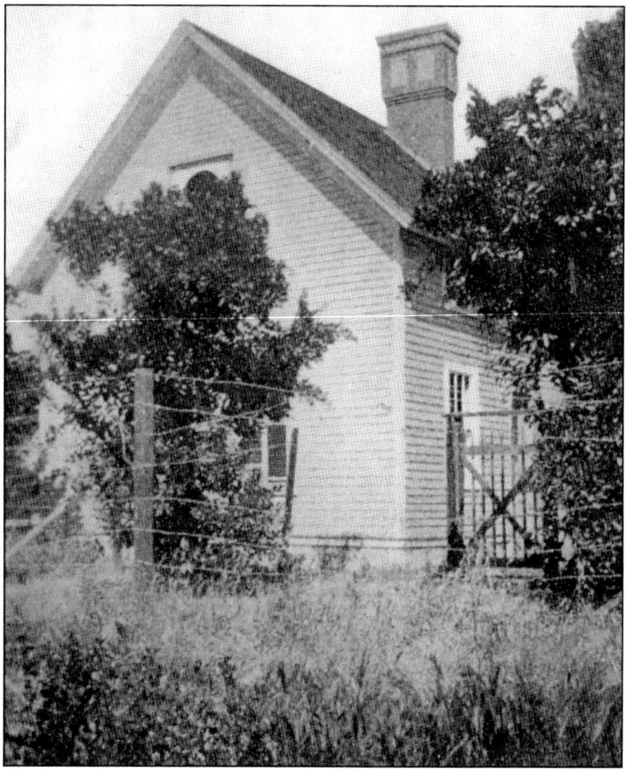

This picture of the Luther and Eliza Tibbets homestead in Riverside, California, shows the two trees sent from the Department of Agriculture surrounded by a wire fence. Eliza Tibbets carefully tended them by hand until they matured. For many years, the Tibbets sold cuttings from their trees, which were originally called the Riverside Navel and later renamed the Washington Navel. This winter-blooming orange provided an incredible boost to the California citrus industry.

Two

Citrus Communities and the Water Wars

Joseph Swift Phillips purchased 2,000 acres of the Rancho La Puente from John Hollenbeck in 1882 and founded Covina, the first citrus community. A self-made man, Phillips worked his way to California on a sailing ship in 1849. He found his gold in dry farming and real estate in Sacramento. Phillips hired engineer Frederick Eaton to divide his land into 10-acre tracts with the town of Covina in the center.

Glendora, the second citrus community, was founded in 1887. George Dexter Whitcomb was a wealthy Chicago industrialist and inventor who came to California for his health. After recuperating for a year, he purchased 200 acres of land at the base of the San Gabriel Mountains and planned a city named Glendora. The word *glen* was for beautiful valley, and *dora* was for his wife, Leadora.

Jonathan Sayre Slauson, a prominent Los Angeles banker, founded the third citrus community, Azusa. In 1880, he acquired part of the Rancho Azusa through a mortgage foreclosure, and in 1884, he brought the remainder of the ranch. In 1886, he sold all but 800 acres to the Azusa Land and Water Company. A year later, the company opened the auction for Azusa lots, selling $280,000 in land on the first day.

The development started a water war over the San Gabriel River, fought by Azusa and Covina settlers. (Glendora sourced its water from the Big Dalton and San Dimas canyons and thus avoided conflict for the most part.) The Azusa Water and Irrigation Company was extending coverage below San Bernardino Road and needed additional funds to complete the system. Phillips bought stock in the company, became superintendent, and completed an irrigation system for Covina. But during the dry years, Azusa settlers felt Covina was getting too much water, and a bitter and vicious conflict began. Both sides appealed to the courts. In 1888, Judge Anson Brunson made a decision allocating the water, but the war did not end.

In 1889, Azusa founder Jonathan Slauson, Judge Charles Silent, and M.L. Wicks of Glendora brought all the parties together, and they signed the Compromise Agreement of 1889 on January 26.

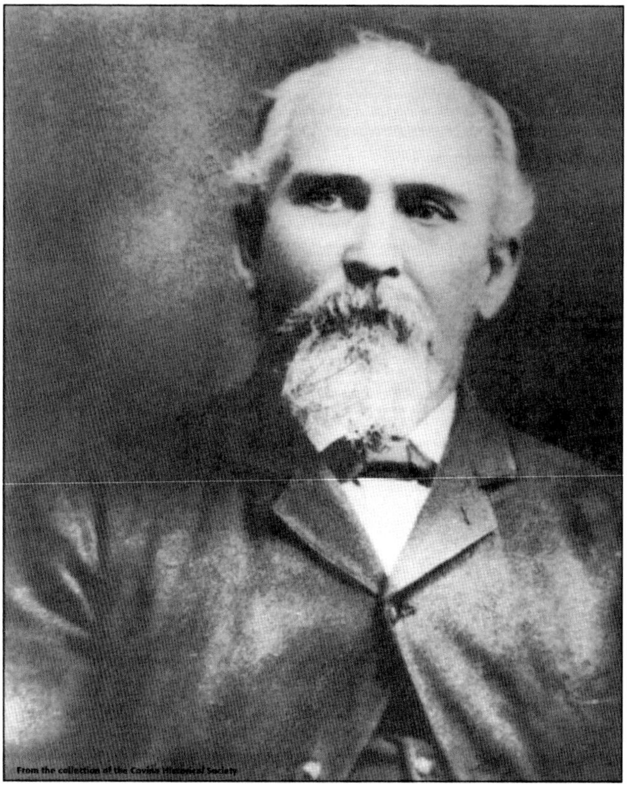

Phillips's land was part of the 4,299 acres Charlotte Gray Rowland had originally sold to the Badillo brothers from Costa Rica. The came to California to start a coffee plantation, but it eventually failed. Joseph Phillips moved into Julian Badillo's former home and pioneered the citrus industry. His vision of a verdant land required water. He spent far more money bringing water to Covina than he invested in the land. He sold Eastern buyers land already planted with citrus.

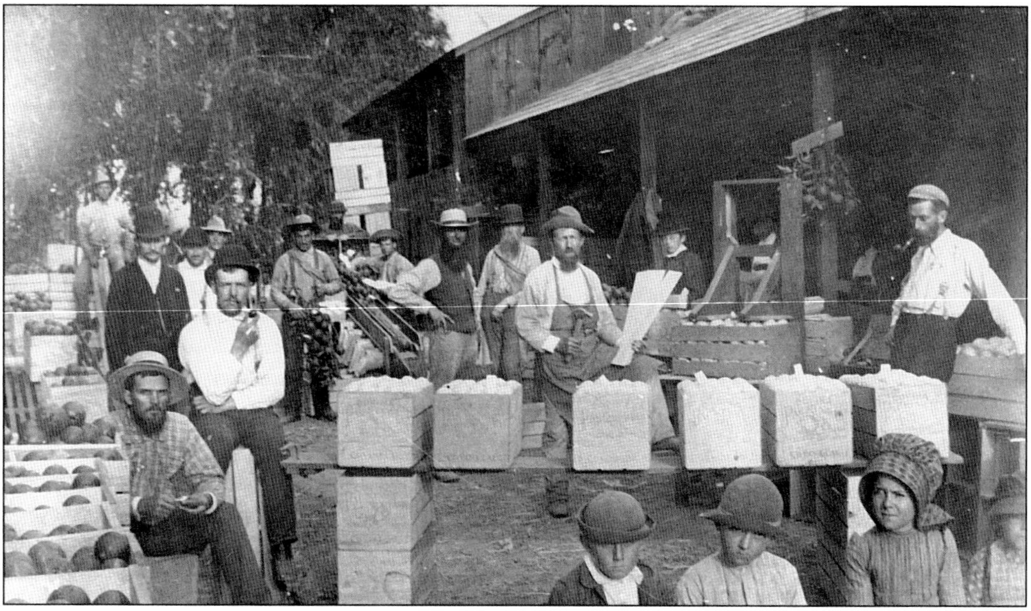

Covina's first commercial citrus operation was the 60-acre Mission Ranch. A mule-drawn wagon took the shipping boxes to the Santa Fe Depot in Glendora. All sale proceeds went to the Church of the Brethren. Daniel Houser established Mission Ranch. He also invented the mechanical harvester and helped finance the ditch that brought water to Covina. This is a photograph of the Mission Ranch packinghouse.

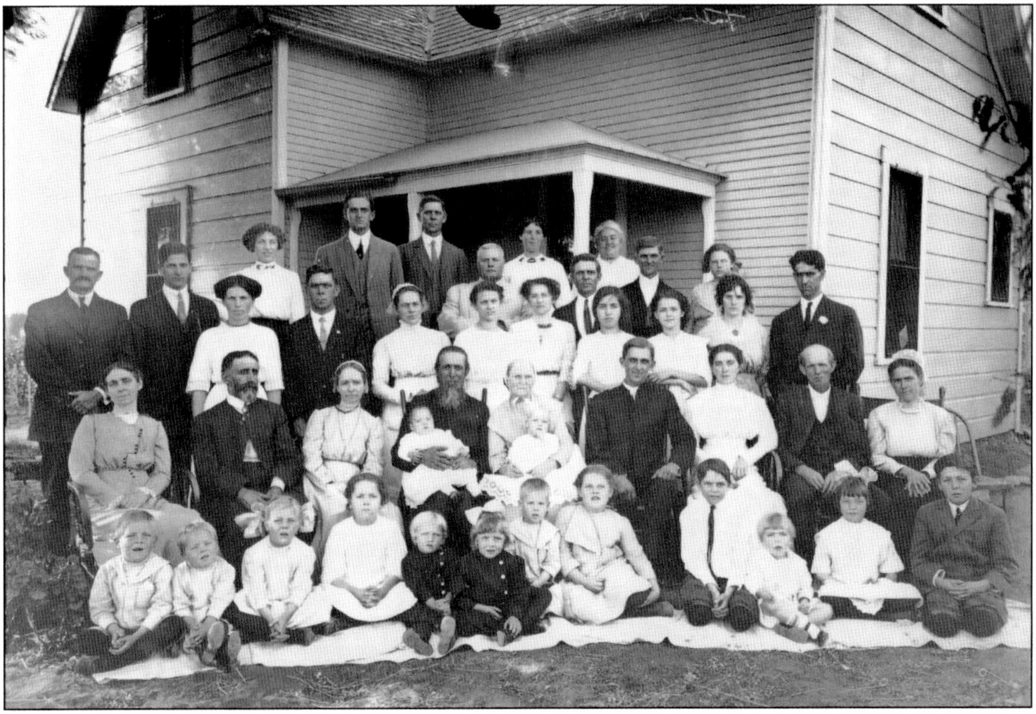

Members of the Zug family, pictured here at a reunion, operated the Mission Ranch for many years. The Brethren completed Covina's first church on January 1, 1887.

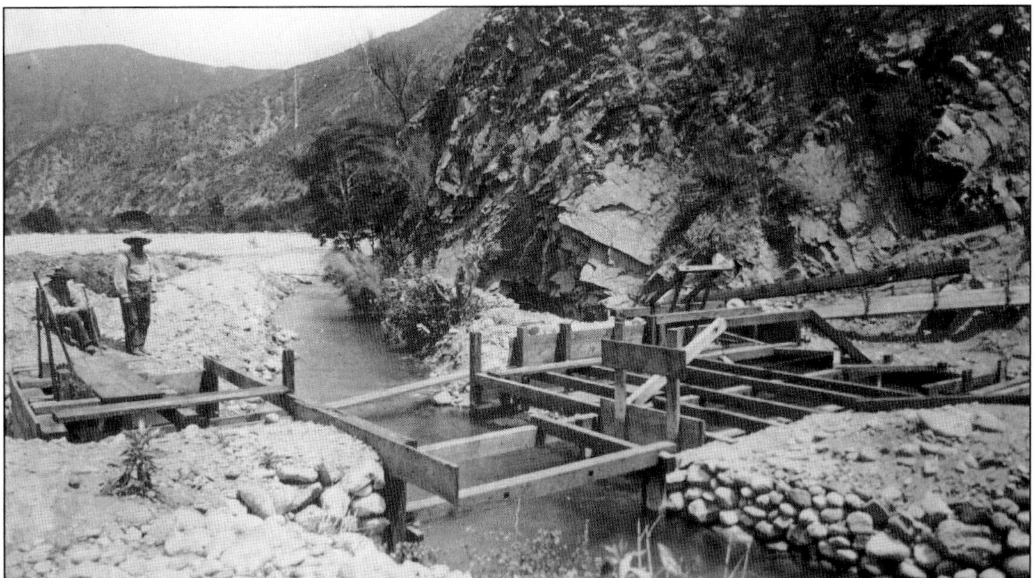

From the collection of the Covina Historical Society

William Griswold recalled in Donald H. Pfueger's book *Covina*, "One time there was a widespread protest when the people of Azusa thought Covina was getting more than its share of water. Every man in Azusa started for the canyon to stop the flow of water. Covina men also rushed to the canyon. Feelings ran high. Whenever we diverted water into our ditch, the Azusa people shut it off. Dams were made and torn out."

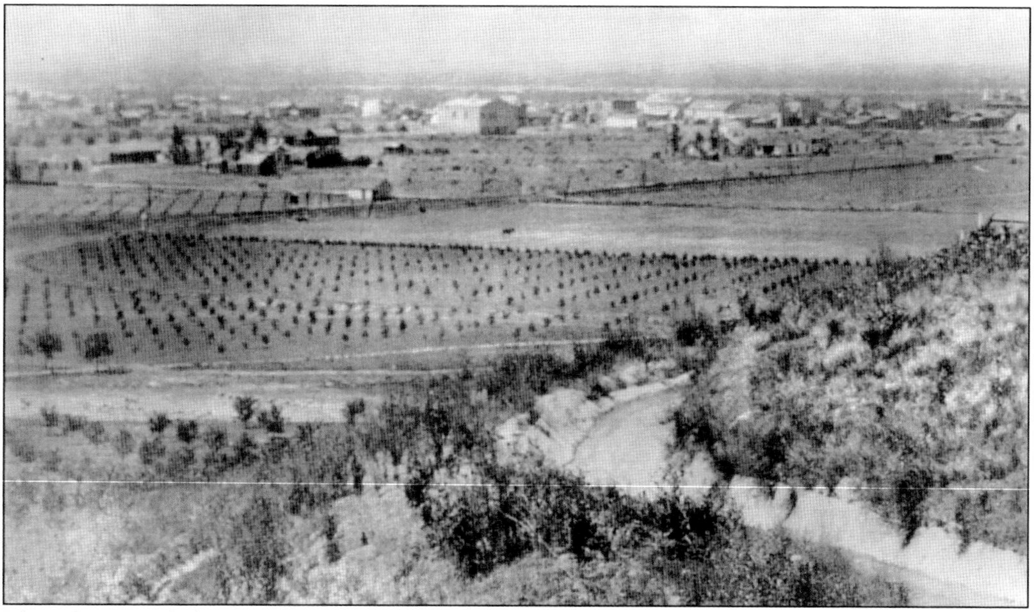

This photograph from 1898 shows the Covina irrigation ditch (often referred to as the Phillips ditch) coming down from the San Gabriel Canyon. W.C. McClintock of Walnut worked for Phillips building the ditch to Covina. V.E. Potter of Covina supervised the cementing of the ditch and reservoir. (Courtesy of Jeffrey Laurence Cornejo Jr.)

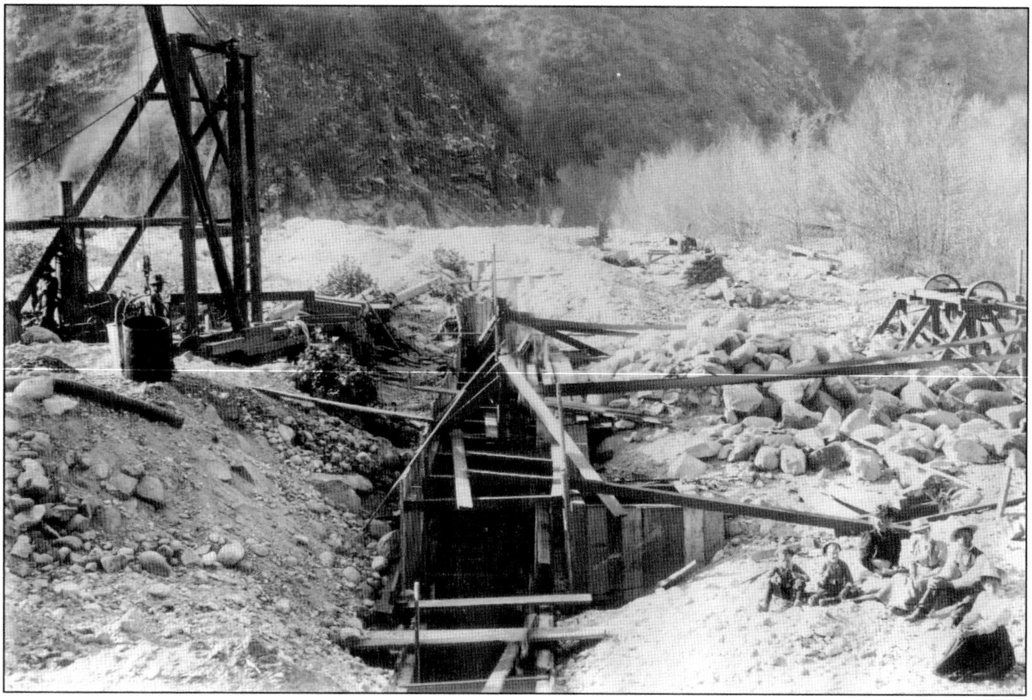

Pictured is a submerged dam built in the San Gabriel Canyon to increase the flow of water to the Phillips ditch. Men from Covina were guarding the tunnel when a group from Azusa destroyed it with dynamite, burying two of the guards. Everyone rushed to dig them out, saving their lives. Although the battle of the San Gabriel River was vicious, no one was killed.

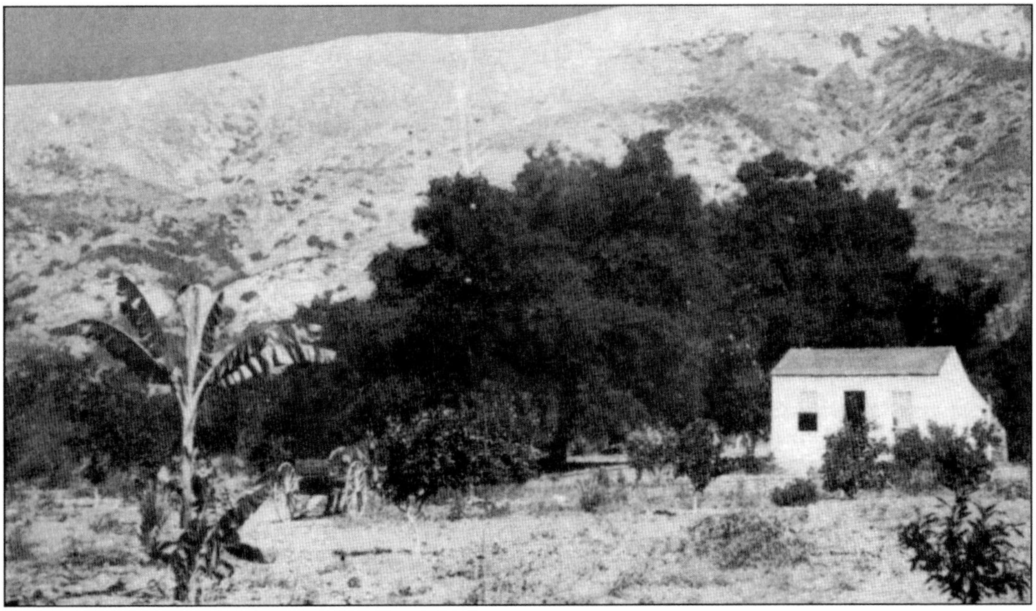

After the government survey that did not recognize Henry Dalton's titles, the land that became Glendora was open to homesteading. The four homesteads on record included John Ganaway's, John Harrar's, Coleman Barnes's, and Leonidas Barnes's. John Bender and William Bryant Cullen bought up promissory rights. The Barnes's homestead between Grand and Pennsylvania Avenues and Foothill Boulevard was purchased by John Bender, whose Springdale Ranch is pictured here. (Courtesy of Glendora Historical Society.)

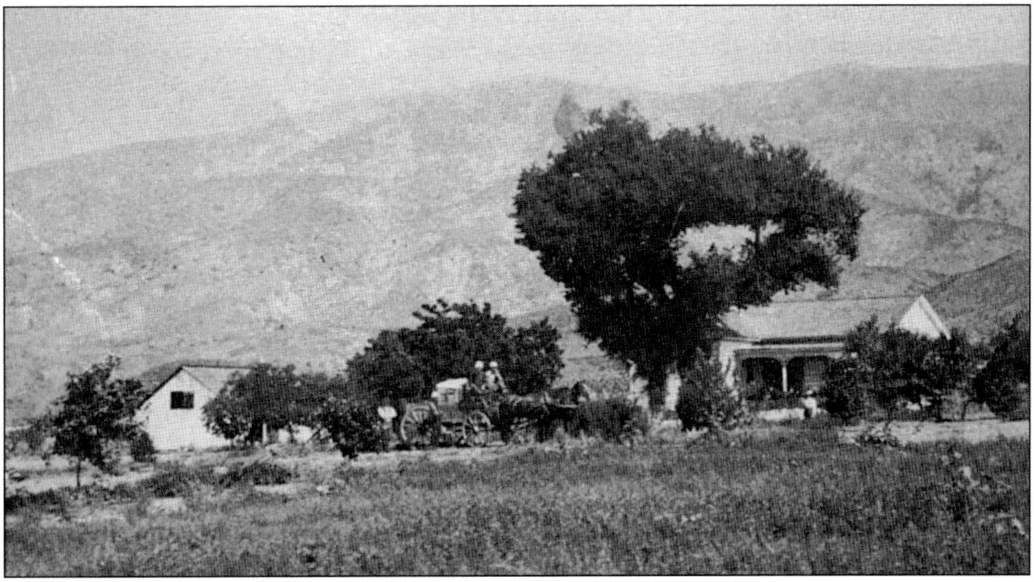

William Cullen's ranch was the original 160-acre Ganaway homestead, bordered by Live Oak and Wabash Avenues to the east and west, and Sierra Madre and Bennett Avenues to the north and south. Cullen's home served as the first post office. The community it served was called Alosta. Bender and Cullen cleared their land with picks and iron bars. As soon as it was cleared, they planted vegetables, grapes, peaches, and seedling citrus trees. Cullen also profited from bees and sheep. (Courtesy of Glendora Historical Society.)

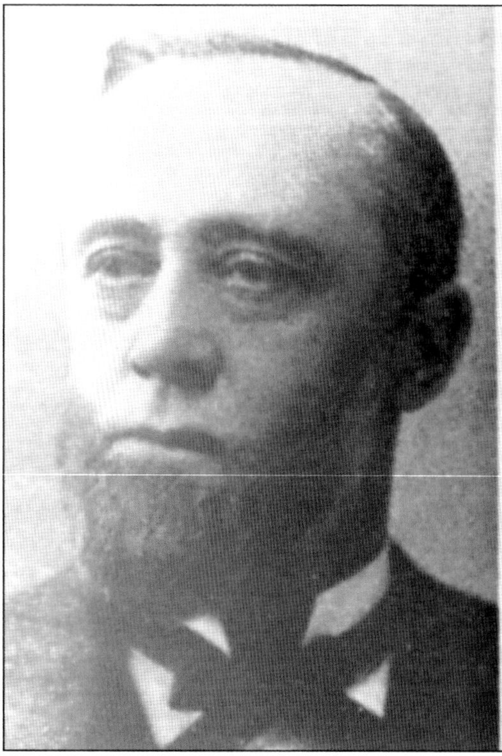

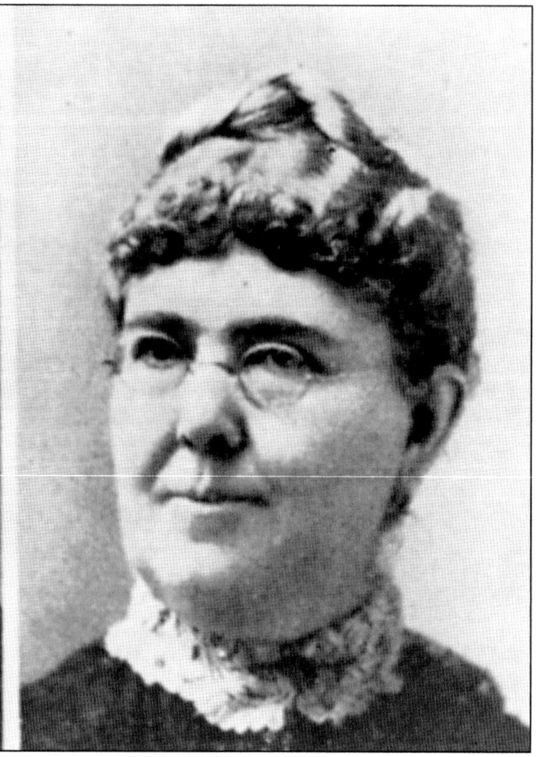

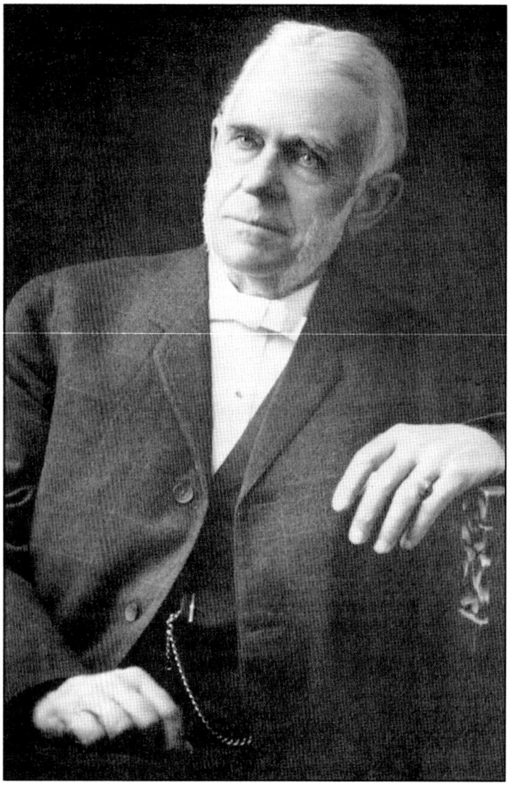

In 1885, George Dexter Whitcomb, pictured with his wife, Leadora, brought a Santa Fe Railroad branch to Azusa and Glendora. John Cook and Merrick Reynolds joined with Whitcomb to form the Glendora Land Company. A land office and the Bellview House, later called the Bellview Hotel, were built. Land was cleared, streets named and graded, and water pipes laid; the opening day was set for April 1, 1887. It is interesting to note that deed restrictions forbid liquor in Glendora. (Courtesy of Glendora Historical Society.)

The Azusa Land and Water Company built the 45-room Azusa Hotel and Slauson's Hall, which served as the city hall, an opera house, a police station, and a public library throughout the years. Slauson's three children, Kate Slauson Vosburg, Louise Slauson Macneil, and James Slauson, created the vast Azusa-Foothill fruit ranch, which included stables, a foreman's house, workers' housing, and a swimming pool. Within a decade, Azusa became a thriving citrus center.

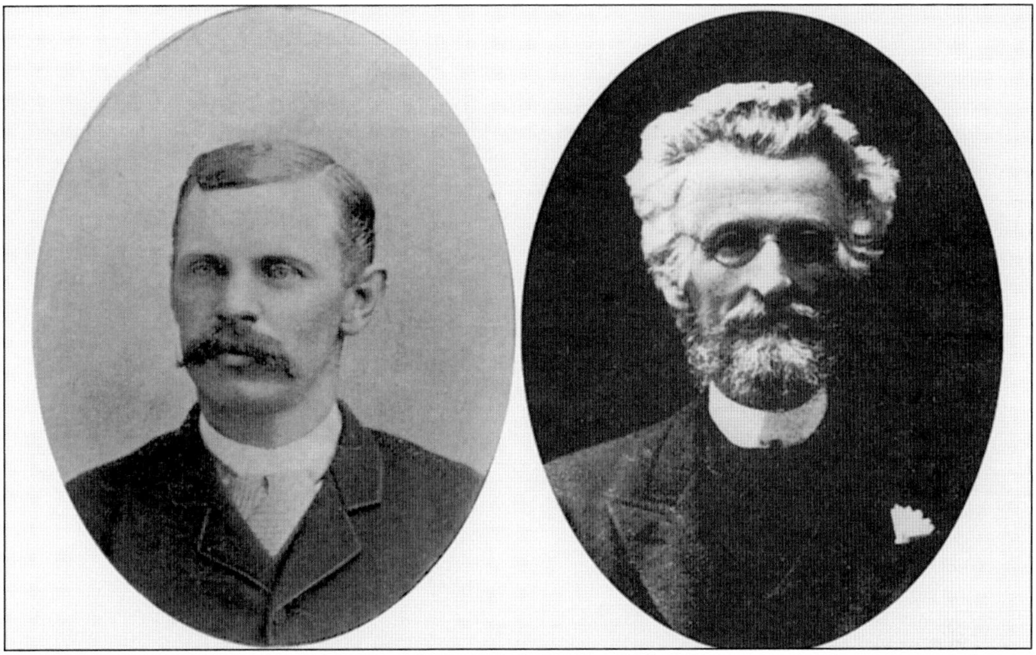

Charles Bemis (left) and Michael Baldridge (right) signed the Compromise Agreement of 1889 for Covina. Bemis was president of the Azusa Water and Irrigating Company (later called Covina Irrigating Company). He helped organize the ACG Citrus Association and became justice of the peace and district agricultural inspector. Michael Baldridge, a pioneer citrus rancher, developed his first grove with seeds from a discarded shipment of oranges.

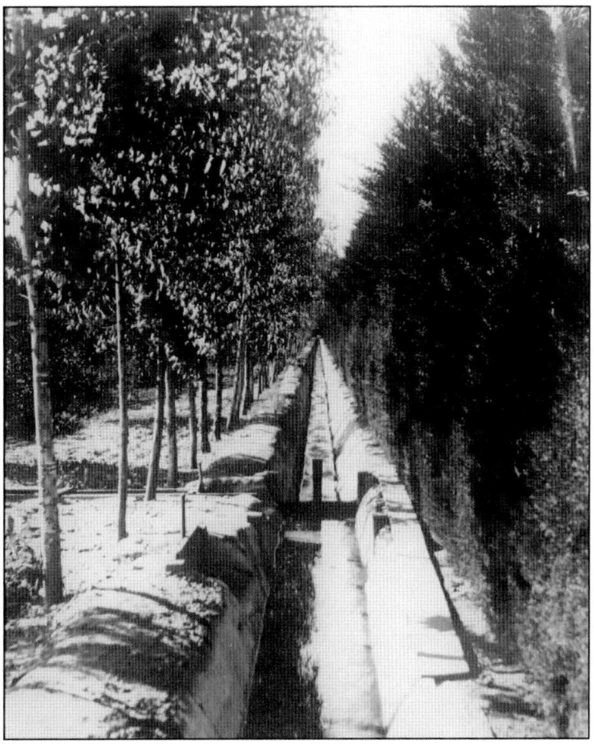

After the water war was over, valley water companies improved and expanded their distribution systems. Main and lateral ditches were created to carry water to the users. Here, a lateral ditch carries water to the Adams Ranch. Stockholders received water according to the number of shares they owned. The person in charge of distribution was known as a "zanjero." James Elliott was the first in the valley to use concrete pipe similar to what is shown.

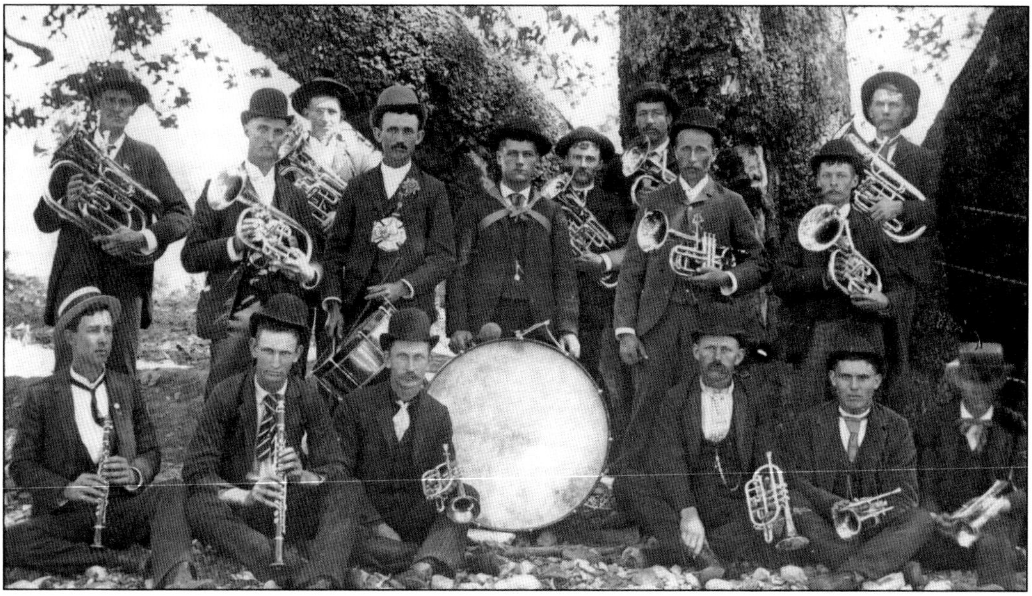

In 1891, Azusa, Covina, and Glendora settlers founded the Oakdale Cemetery. The Covina Cornet Band played at a Memorial Day picnic at the cemetery. Pictured from left to right are (first row) Carl Warner, Eugene Griswold, William Griswold, H.F. Headley, John King, and William Walker; (second row) Frank Cartzdafer, Marion St. Clair, Harvey Preston, E.G. Prather, Pat Talley, Charles Earle, Arthur Eckles, John Remalex, Arthur St. Clair, and Harry Amon.

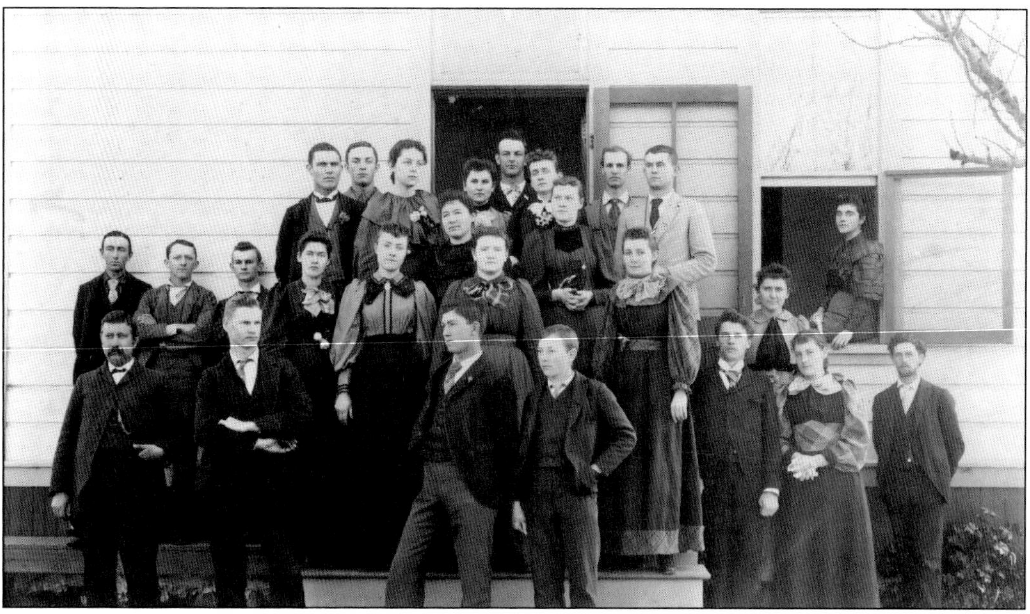

Citrus Union High School students are pictured from left to right as follows: (first row) principal Major Meredith, Herman Lee, George Sunt, and unidentified; (second row) unidentified, T. Rogers, Homer Hosteller, Minnie Pierce, Ida Plummer, unidentified, Ina Reeves, Robert Owens, Emily Hudson, Zada Taylor, and George Moxley; (third row) John King, Dick Guiberson, Zuleika Guiberson, and Clara Eckles; (fourth row) May Griswold, Nat Guiberson, Gertrude Vaughn, Helen Clapp, Gifford See, and teacher Isabelle Owens, seated in the window.

Three

First Citrus Cooperative and Pioneer Nurserymen

In 1882, Joseph Swift Phillips started dry farming grain on his 2,000 acres. He used a new harvester, designed by Daniel Houser, which cut, threshed, and sacked the wheat for shipping.

As water gradually became available, small farms replaced the sea of grain with fruit trees, vegetables, and berries. James Hodges planted a Salaway peach and a French plum, the first fruit trees in Covina. He and the other pioneer nurserymen in the valley cleared cactus, large sunflowers, and caster beans to raise seedlings. During dry periods, they brought barrels of water from the river to irrigate the trees by hand. The Covina Valley citrus industry began with James Hodges, W.R. Barbour, Michael Baldridge, J.R. King, A.L. Keim, John Coolman, and Madison Bashor. At first, they grew both deciduous fruit trees and citrus. As the market for citrus grew, citrus-bearing trees almost entirely replaced the deciduous fruit trees.

The *Glendora Signal* from May 1888 included an advertisement that read, "Non-residents look HERE. Orange growers read. Best offer yet made. W.R. Barbour is prepared to furnish orange trees in any variety at a uniform price of $1.50 per tree and insures for three years." The advertisement goes on to explain the terms of the deal: Customers would pay 50¢ per tree in the first year and 25¢ for the next two years.

By 1889, valley ranchers had another source of citrus stock in San Dimas. Robert Teague started his one-acre nursery with 10,000 trees. By 1907, he had 700,000 seedbed-stock seedlings growing in the open, making Teague's operation the largest citrus nursery in the world at the time.

Inspired by the organization of the Southern California Fruit Exchange, Covina Valley ranchers formed the ACG Citrus Association at Slauson's Hall in Azusa in 1893. According to the *Azusa Pomotropic*, W.R. Barbour proposed that the Azusa, Covina, Glendora Citrus Association organize with an incorporated stock of $10,000 and that each grower who wished to join the association would have to buy one share of stock for each bearing acre. C.E. Bemis made the incorporation list. The *Azusa Pomotropic* also reported that the growers succeeded in subscribing "to the full extent of their bearing acreage." The ACG directors met with the Pomona and Ontario Exchanges and formed the San Antonio Fruit Exchange.

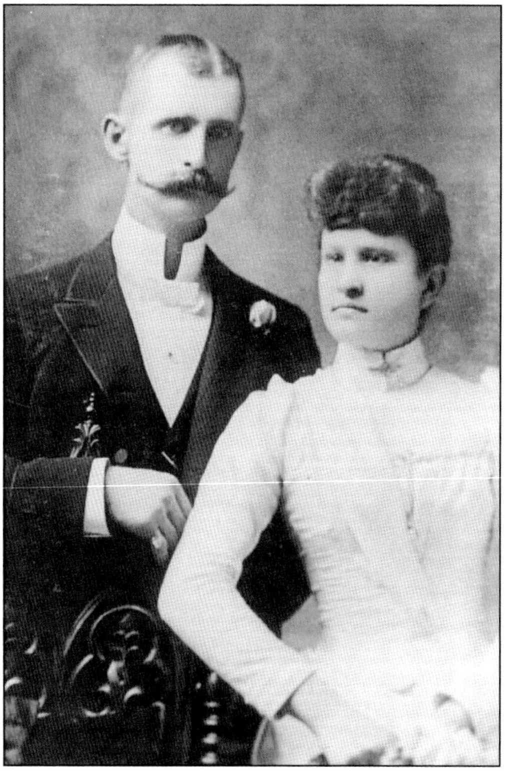

This photograph documents the marriage of Aron L. Keim to Estella Judd. Pastor James Crawford married the couple at the Presbyterian church in Ventura on February 16, 1891. Keim was a successful young Covina businessman who established one of the first nurseries in the valley. An amusing story from the *Azusa Pomotropic* tells how Keim had to appease his bachelor friends with two boxes of cigars after his wedding. (Courtesy of Karen Shomber.)

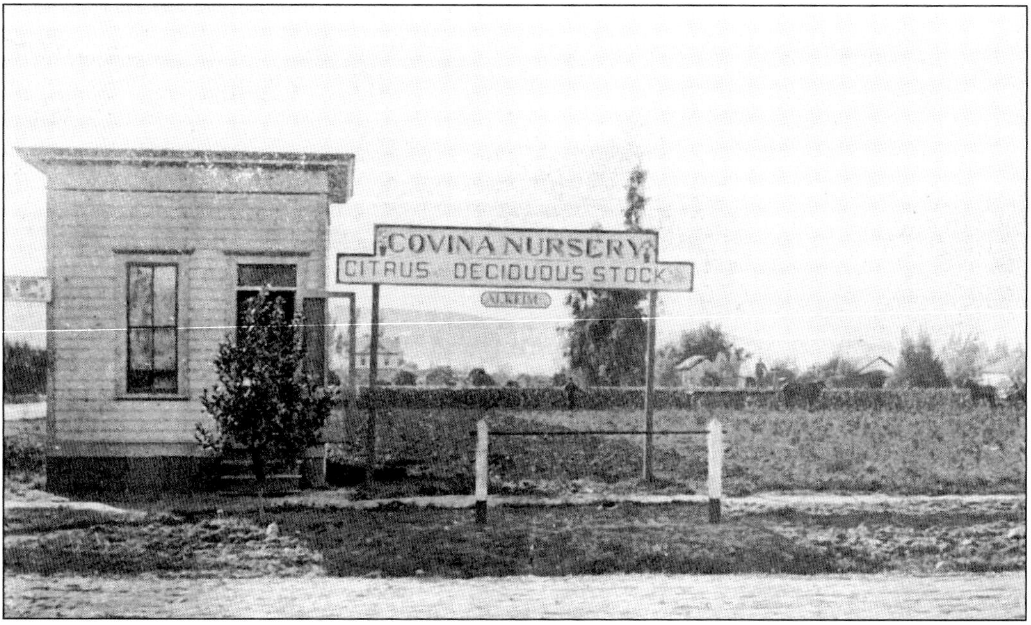

Aron L. Keim's nursery, pictured, became one of Los Angeles County's largest citrus suppliers. Keim propagated Thompson Improved and Washington Navel oranges, Valencia late oranges, and Eureka, Lisbon, and Villa Franca lemons. In addition to citrus, he grew deciduous fruit, including peaches, apricots, plums, and prunes. He also owned a real estate business specializing in Azusa Valley properties.

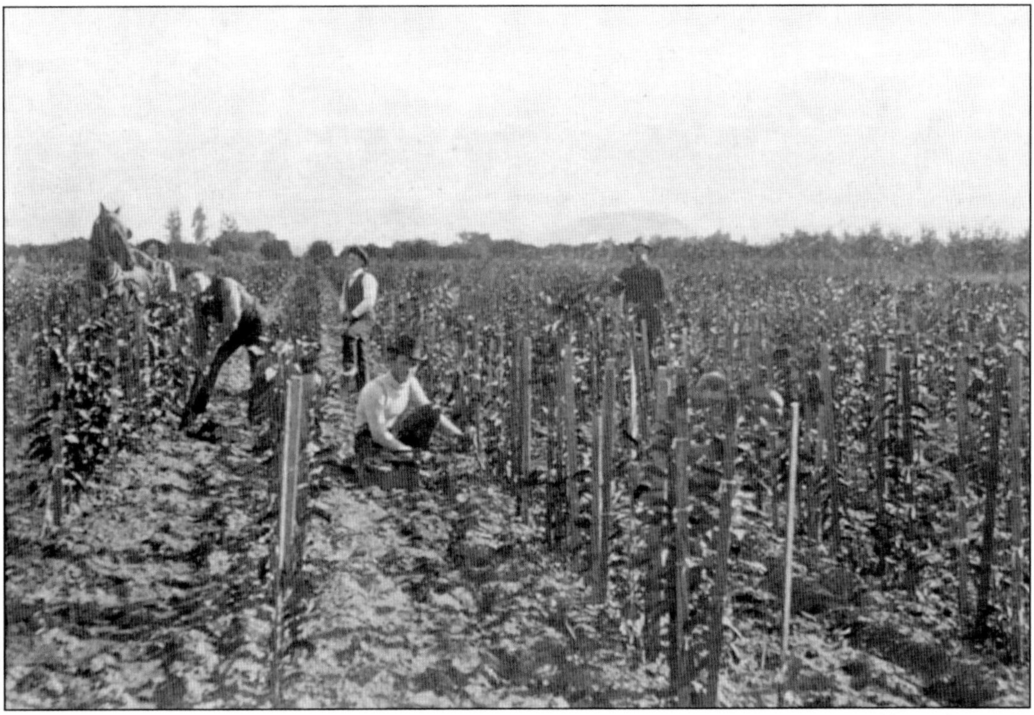

The King nurseries pictured were founded by John R. King in 1885. The nurseries sold Washington Navel and Valencia orange trees. King's specialty was the Thompson Improved Navel, which he budded from his own stock. He purchased his first Thompson Improved trees from their originator, A.C. Thompson of Duarte. King was a prominent member of the Independent Order of Odd Fellows and active in the Ancient Order of United Workman.

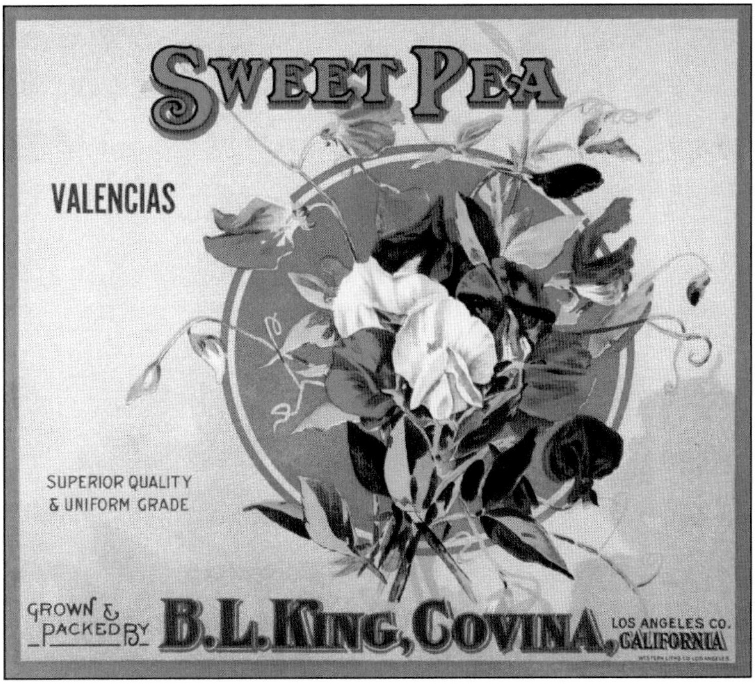

Boyd King was one of the first citrus ranchers to grow, pack, and ship his own fruit. Pictured here is Sweet Pea, one of his early citrus crate labels, printed by Los Angeles Western Lithograph Company.

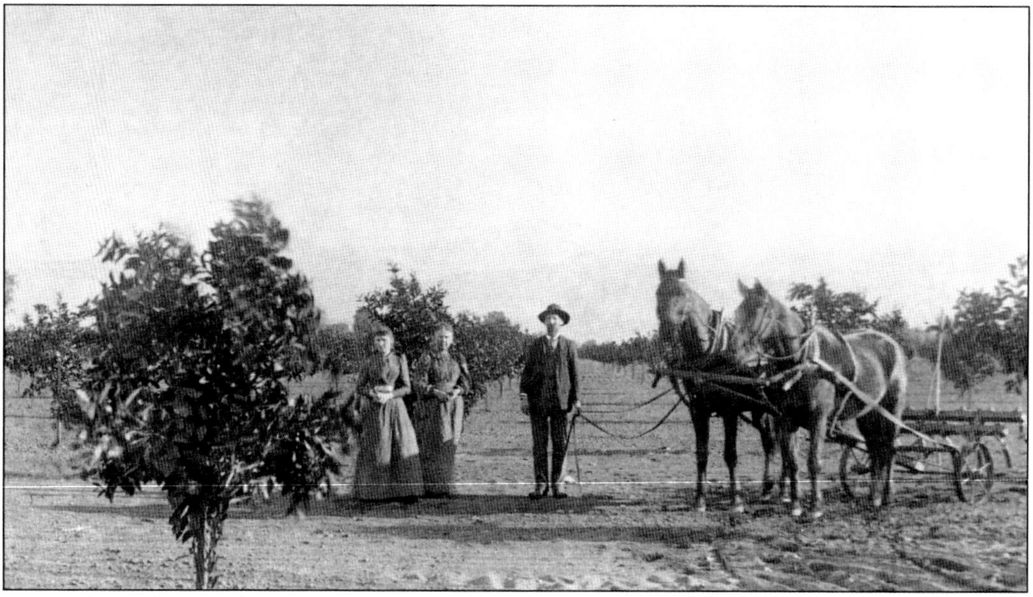

Pictured standing in their 23-acre orange grove (presently downtown Covina) are, from left to right, Frances Cook, Italia Richmond Cook, and Benjamin Franklin Cook. In 1886, Italia arrived from Iowa to visit her father, a shopkeeper on Badillo Street. She had left negative-20-degree weather in Iowa, and the climate and the little village so impressed her that Italia persuaded her husband, Benjamin, to move to Covina.

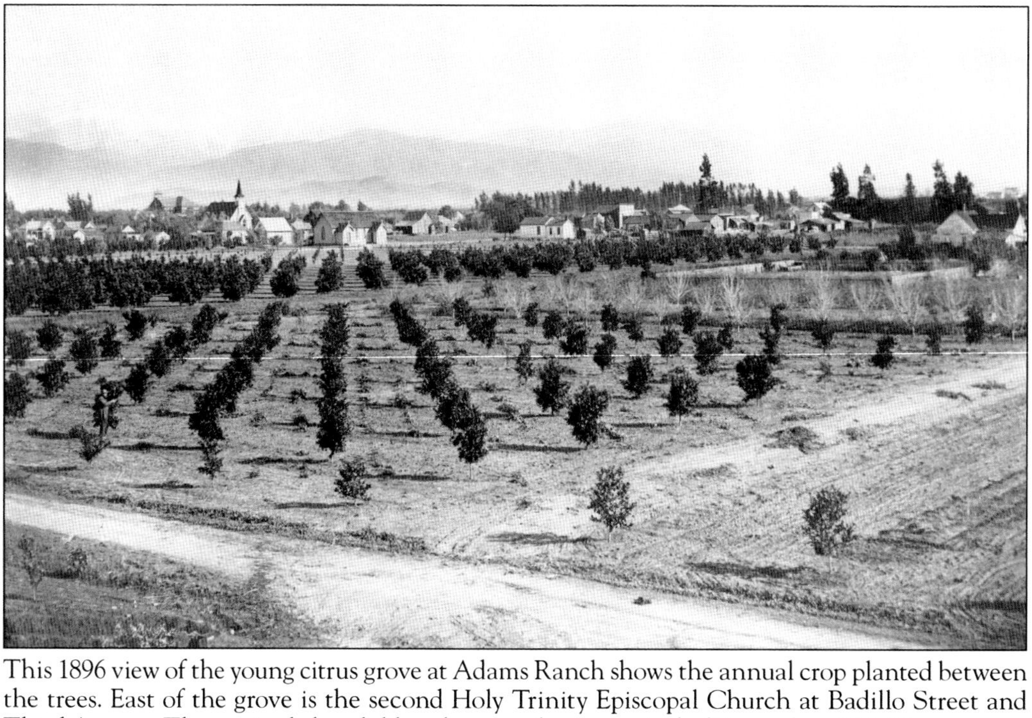

This 1896 view of the young citrus grove at Adams Ranch shows the annual crop planted between the trees. East of the grove is the second Holy Trinity Episcopal Church at Badillo Street and Third Avenue. The original church blew down in the great wind of December 1891. The growing village can be seen beyond the church. Farmers' families came to shop on Saturday nights, and everyone was invited to dances in the social hall.

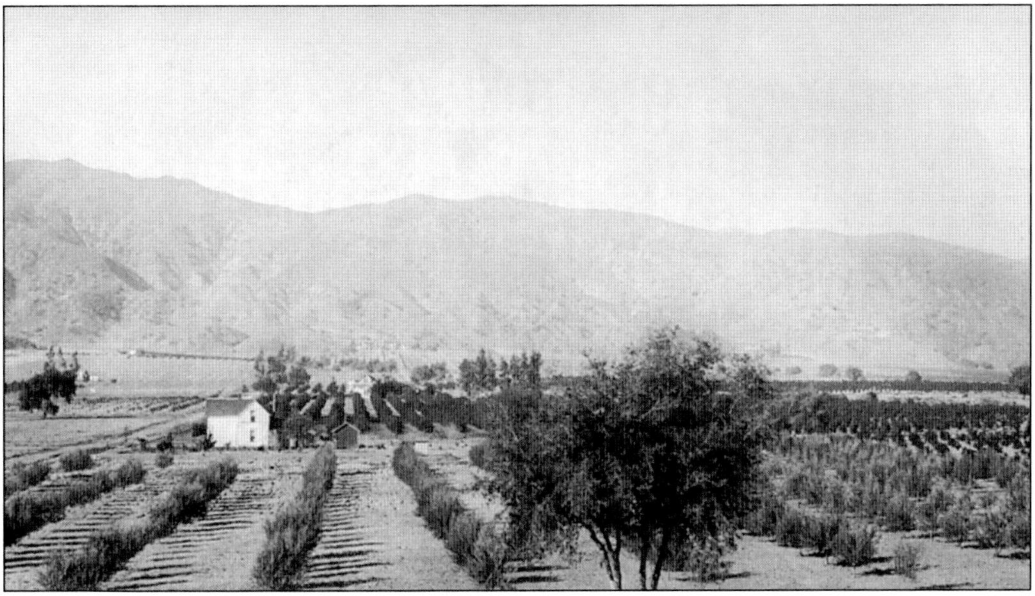

Pictured is the Glendora peach and citrus grove of J.J. Wamsey. Behind the Wamsey grove on the left side of the photograph is the 10-acre navel orange grove planted by John W. Cook in 1886, said to be the first commercial Washington Navel orange grove in the upper Azusa Valley. Cook became associated with Whitcomb through the Glendora Land Company. (Courtesy of Glendora Historical Society.)

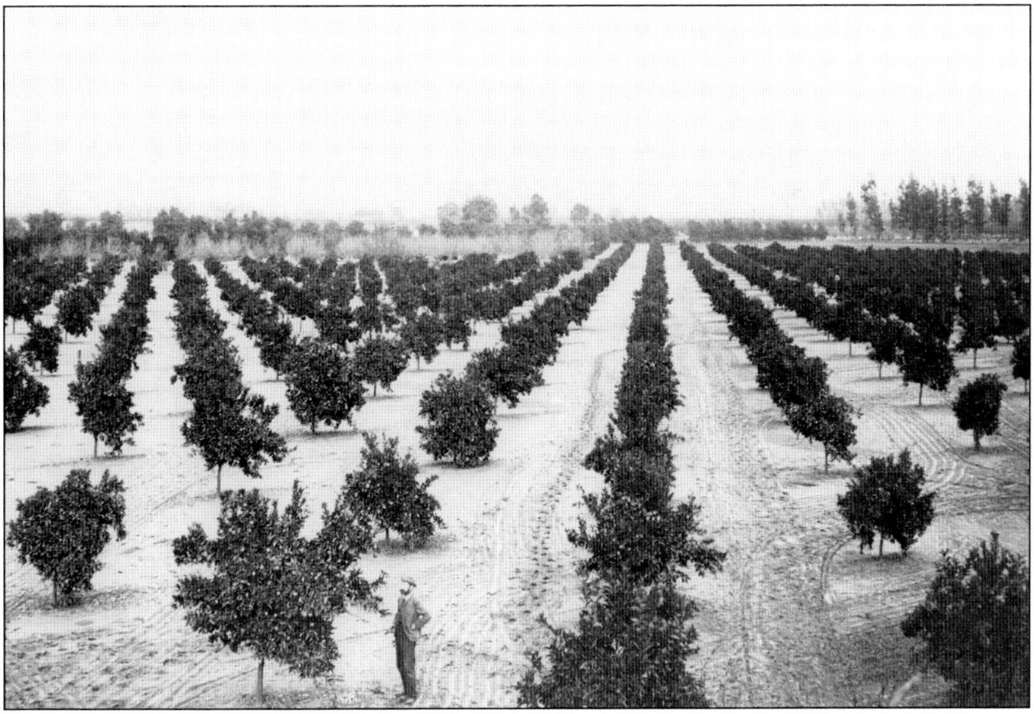

Pictured is an early citrus grove in the Azusa area. Henry Roberts, who lived near the mouth of the San Gabriel River, was the first citrus packer and shipper in Azusa. He employed local workers and returned much of his profit to the growers.

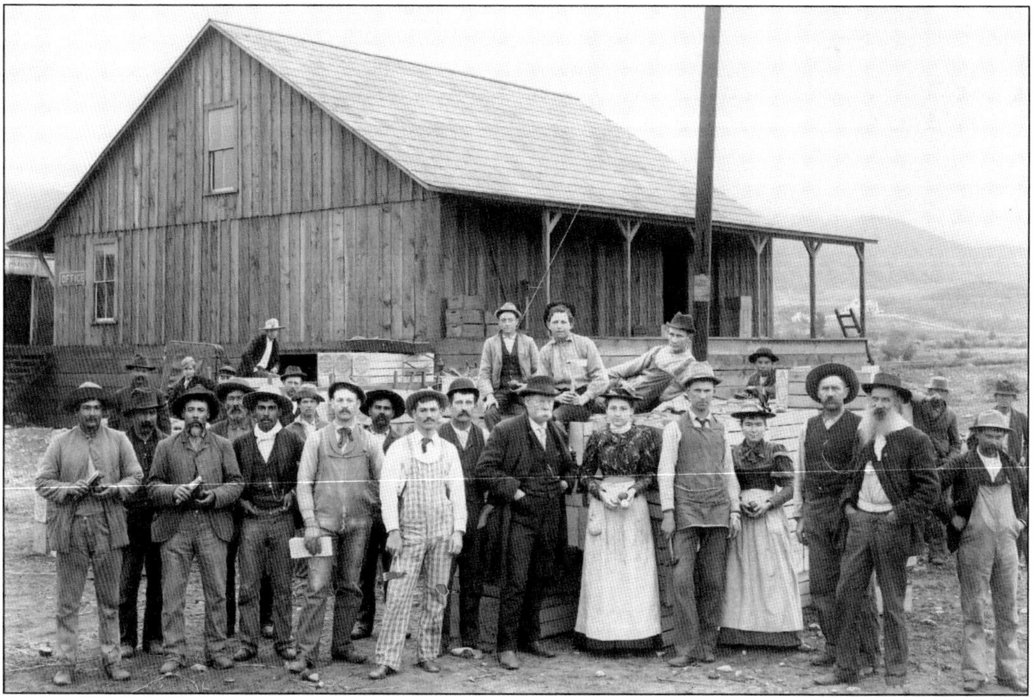

This is the only known exterior photograph of the first ACG Citrus Association packinghouses, which were built adjacent to the Santa Fe Railroad in Azusa and Glendora. This one was in Azusa. Crate makers, washers, and packers are standing with the manager, G.W. Taylor, sixth from the left in the first row. On his left is Percy Jackson. Seated behind him leaning against the pole is Eugene Griswold, son of Thomas Griswold. (Courtesy of the Elliott-Milliken Collection.)

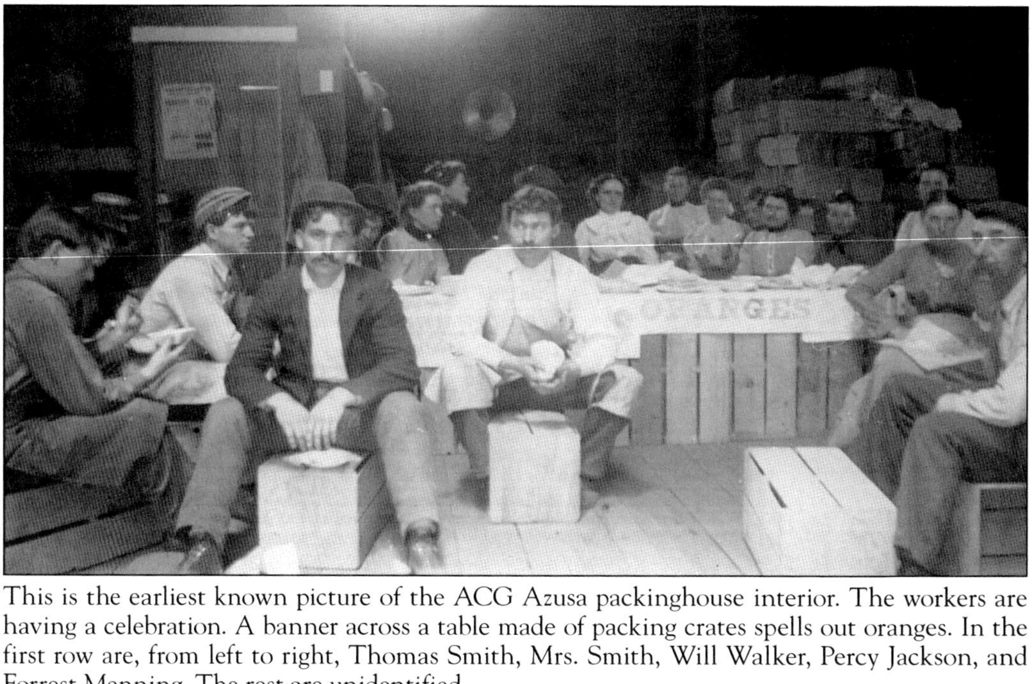

This is the earliest known picture of the ACG Azusa packinghouse interior. The workers are having a celebration. A banner across a table made of packing crates spells out oranges. In the first row are, from left to right, Thomas Smith, Mrs. Smith, Will Walker, Percy Jackson, and Forrest Manning. The rest are unidentified.

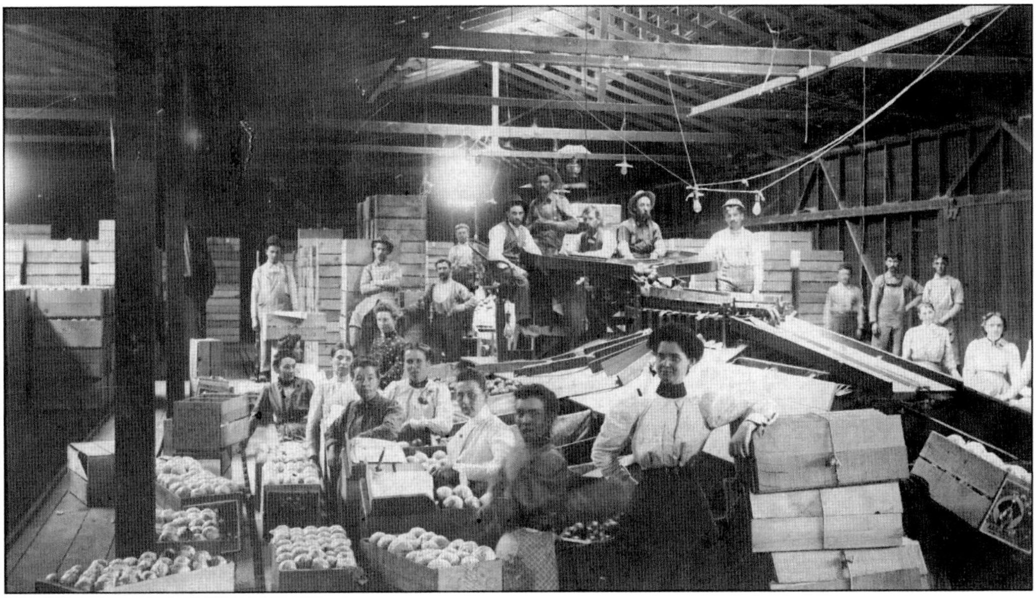

Shown is the interior of the ACG packinghouse in Azusa in 1900. Packing crates are marked with Pointer labels. The lady packers on the left side of the photograph, from front to back, include Betsy Barlow Jackson, Mrs. Ed Dorsay, Mrs. Thomas Smith, Mrs. Lena Williams, Mrs. Stienke, Carrie Burns, Mrs. Lovelace, and two unidentified women. Behind the sorting machine are, from left to right, three unidentified men, Art St. Clair, Thomas Smith, Percy Jackson, Arnold ?, Will Walker, and Charles Gable. The rest are unidentified.

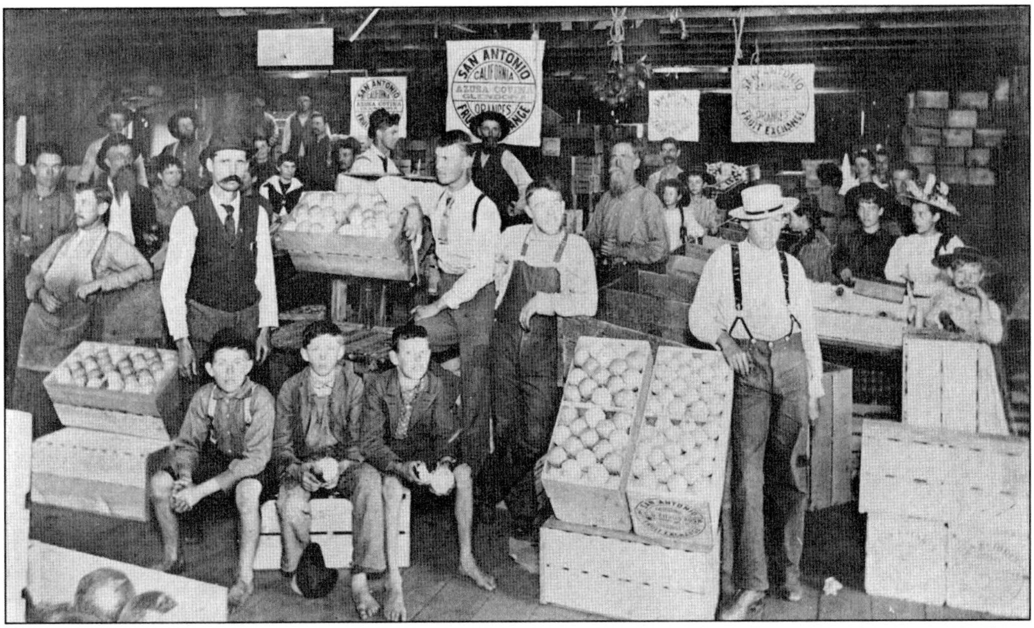

This rare interior picture of the first ACG packinghouse in Glendora shows employees with their families attending a celebration. Notice the circular San Antonio Fruit Exchange label stamped at the end of the packing crates. The medallion also appears on the banners hanging in the back. The packinghouse was located near the Santa Fe Railroad tracks, but its exact location is unknown.

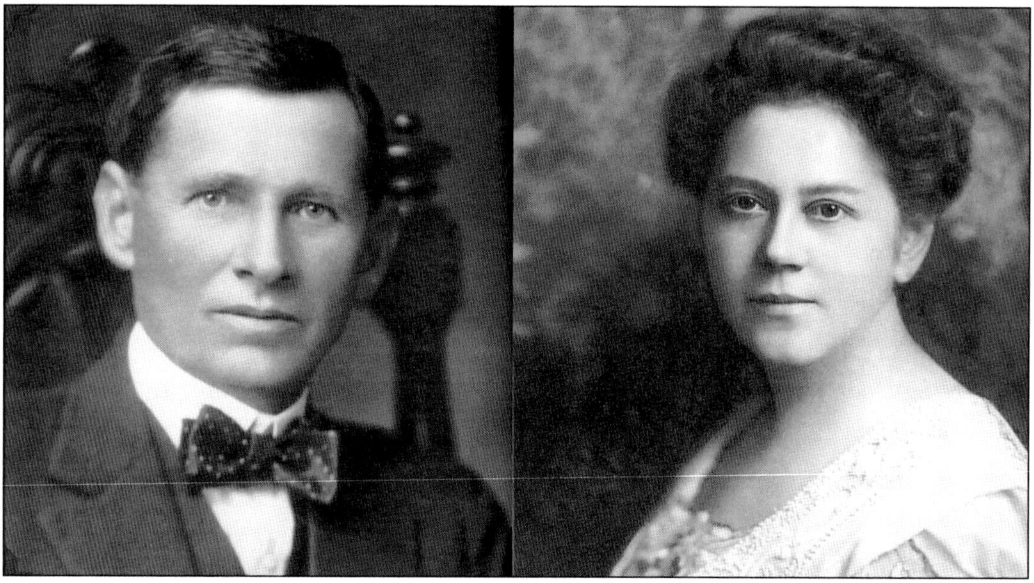

In 1882, Anton P. Kerckhoff purchased 40 acres east of Covina; he planted 25 acres of navel oranges and 15 acres of lemons. After Kerckhoff and Frances Porter (pictured) were married, they lived with his parents in Los Angeles while their home was being built. They checked in on the building by taking the Santa Fe Railroad to Glendora and riding in a buggy to the site. (Courtesy of Alice and Jerome Fahey.)

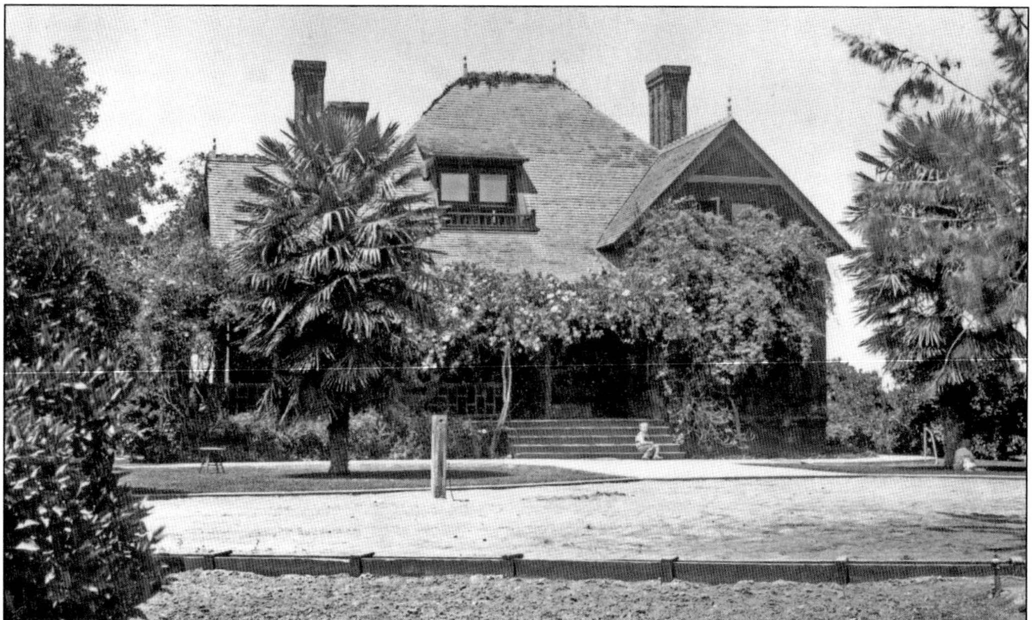

Anton named their home La Francesquita for his bride. Oak trees and beautiful gardens with specimen plants, including a giant Abyssinian banana tree, surrounded the Elizabethan-style home. The driveway was lined with palms and roses. Knowing that the valley citrus industry needed more rail transportation, Anton and his brother William worked tirelessly to bring the Southern Pacific to Covina.

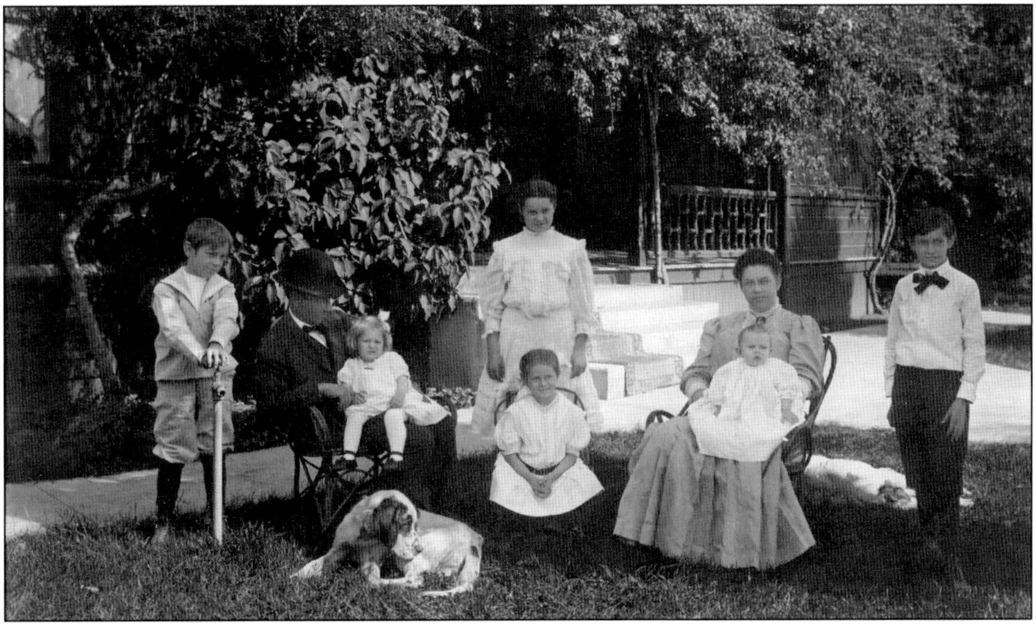

The Kerckhoff family is pictured from left to right as follows: Porter, Anton Kerckhoff holding young Frances, Rosita, Elise, Frances Kerckhoff holding baby George, and Philip. Anton Kerckhoff strongly supported cooperative development of the valley citrus industry. He served as president of the Covina Citrus Association, the Covina Irrigating Company, and the Covina Fruit Exchange. He and Claiborne Vaughn are credited with saving the ACG Lemon Association. (Courtesy of Alice and Jerome Fahey.)

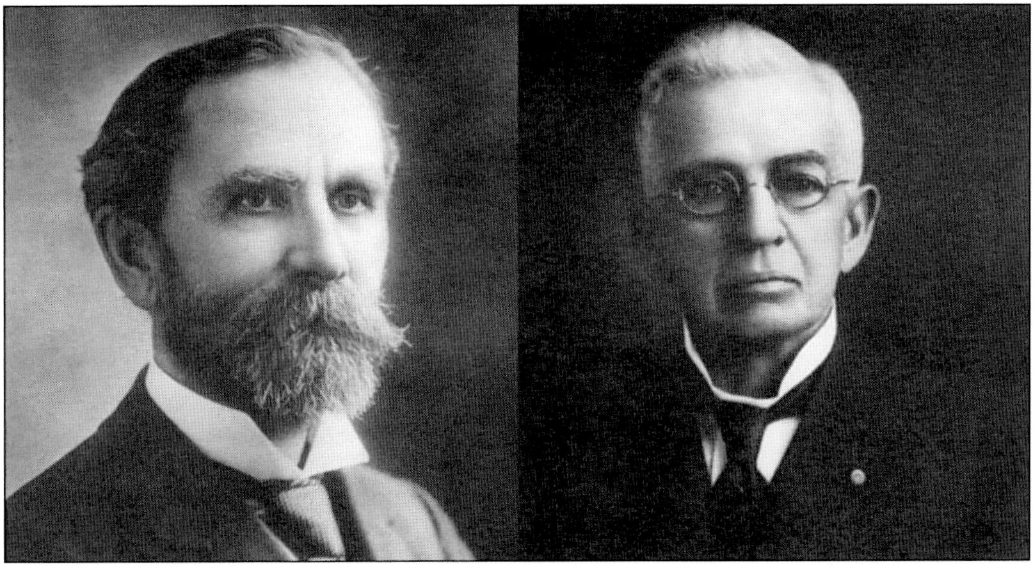

Pictured on the left is T.H.B. Chamblin and on the right is P.J. Dreher. Chamblin is considered the father of the Southern California Fruit Exchange. In 1892, he and 11 of his Riverside neighbors organized the Pachappa Orange Growers Association. He presented his idea for the Southern California Fruit Exchange at a citrus ranchers meeting in Los Angeles, and they agreed to start the exchange. Chamblin and Dreher personally visited as many citrus ranchers in Southern California as they could to urge them to join.

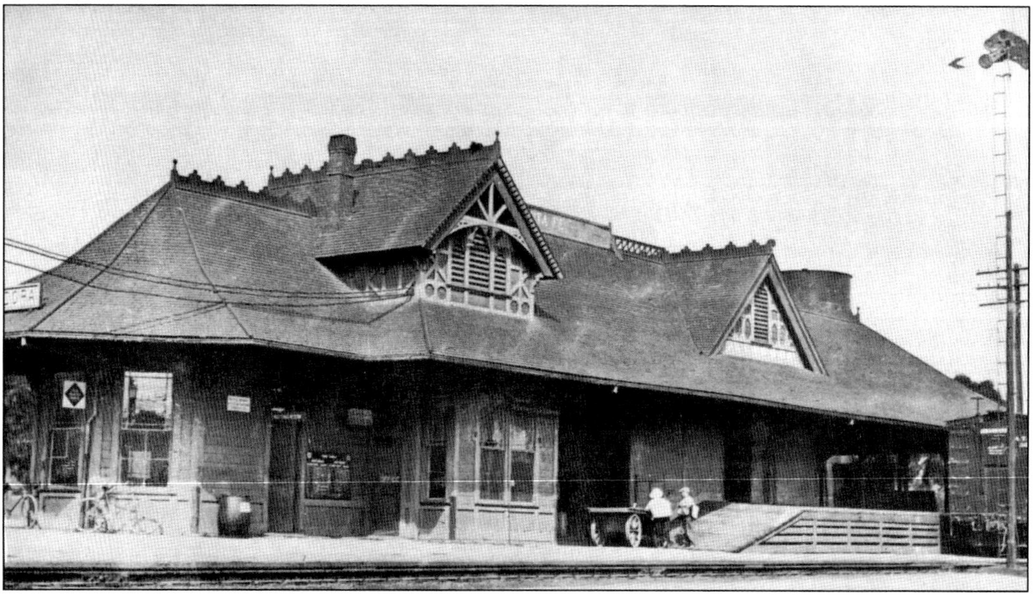

Pictured is the Glendora Santa Fe Railroad station, which was located in the community formerly called Alosta. By March 1887, two trains arrived daily. They stopped in Azusa and Glendora and went on to San Bernardino. For the first time, citrus ranchers had real access to Midwestern and Eastern markets. (Courtesy of Glendora Historical Society.)

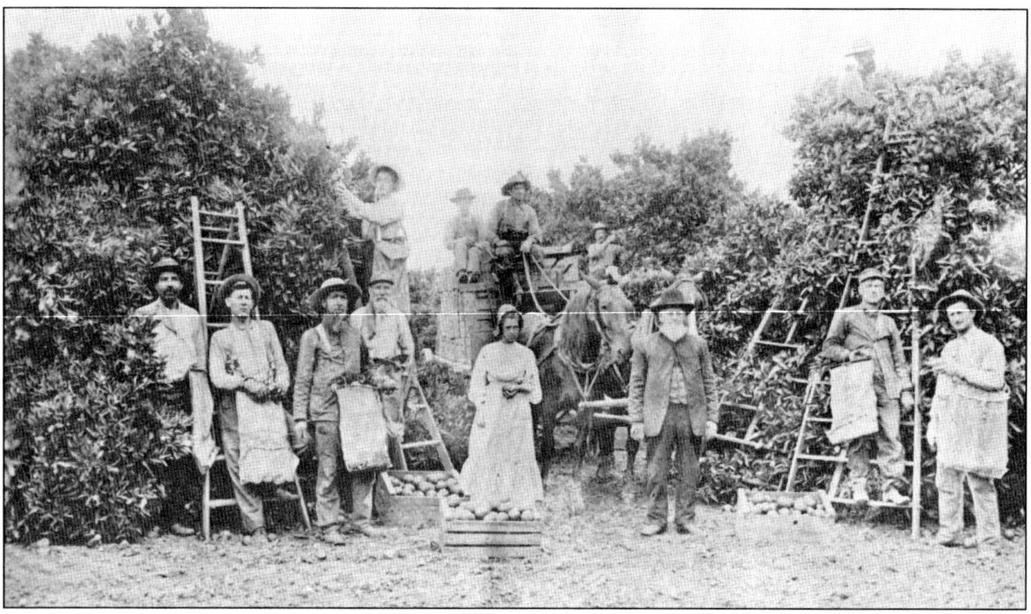

John and Sylvia Netzley are standing in the center of this photograph. They managed a large citrus ranch near present-day Glendora and Mauna Loa Avenues. Photographs showing citrus pickers can be divided into three period styles. This one is from the second period, which showed owners or managers together with actual pickers. Owners were often well dressed and riding in a buggy or an automobile. In later years, Sylvia Netzley became well known for her weaving.

Four

New Citrus Cooperatives and Independent Packers

John Coolman and Joe Clarke led a vigorous campaign to persuade Henry Huntington to bring the Southern Pacific Railroad through Covina. Their committee of civic leaders, which included Anton and William Kerckhoff, secured the necessary right-of-ways, and the railroad arrived. The hub of citrus shipping in Covina Valley later shifted south along the Southern Pacific tracks that ran from Irwindale through Covina and Charter Oak. A symbiotic relationship developed between the valley citrus industry and the real estate promotion. Valley newspapers joined the effort. The *Covina Argus* published glossy holiday magazines in 1896 and 1898. They featured homes surrounded by beautiful citrus groves. They ranged from Roseland, the home of retired Chicago railroad conductor George Anderson, to the luxurious ranch of J.H. Adams, inventor of the Pullman railroad car. An advertisement from the *Covina Argus* in 1898 reads:

> Come to Southern California and make your home among the luxurious groves of Covina, clothed in their emerald mantle and laden with their golden fruit, the very atmosphere laden with the scent of their blossoms. Covina, situated on the Southern Pacific Railroad, nestled in the very center of the most luxurious and productive valley in Southern California, the San Gabriel, sheltered from the heat of the desert during the summer and the cold north blast in the winter, by the great chain of the Sierra Madre Mountains, whose lofty peaks stand out grim and stern against the azure sky, with deep crevasses that seam their granite slopes and their rugged rocks half hidden by mosses and trickling vines, their sharp angles smoothed by welcome streams of water (caused by the melting snow accumulated during the winter) that dash from their dizzy heights, while here and there at lower altitudes the spreading branches of mighty trees cast their shadows and delicate flowers nestle at their base. The orange is only one of the many products which contribute to the happiness and prosperity of our ranchers. It is the place for those who desire the serenity of rural life combined with the culture and advantages that promote intelligence and refinement.

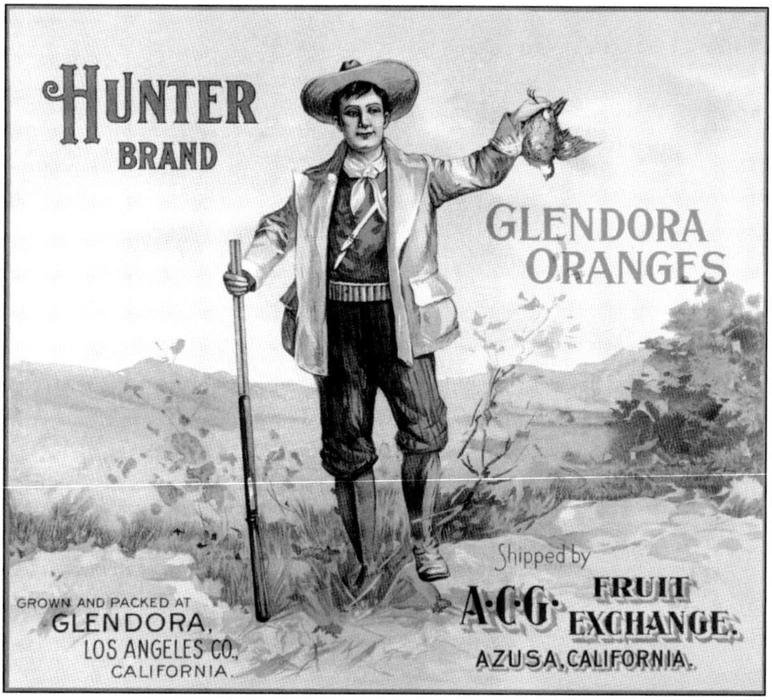

The Hunter Brand Glendora label was designed and printed by the Stecher Company in Rochelle, New York. The oranges were packed and shipped from Glendora through the ACG Fruit Exchange in Azusa, California. The hunter and the bird he is holding are not realistic portrayals of Glendora hunters or the game birds they shot, which were mostly quail and dove.

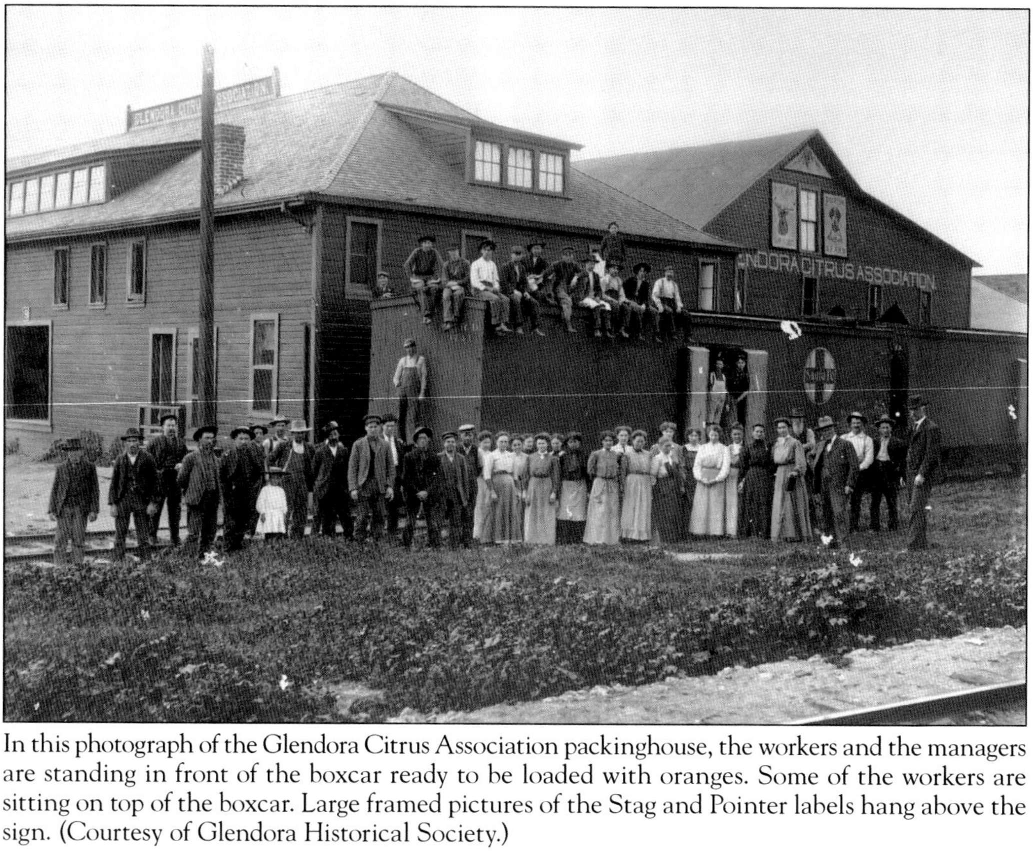

In this photograph of the Glendora Citrus Association packinghouse, the workers and the managers are standing in front of the boxcar ready to be loaded with oranges. Some of the workers are sitting on top of the boxcar. Large framed pictures of the Stag and Pointer labels hang above the sign. (Courtesy of Glendora Historical Society.)

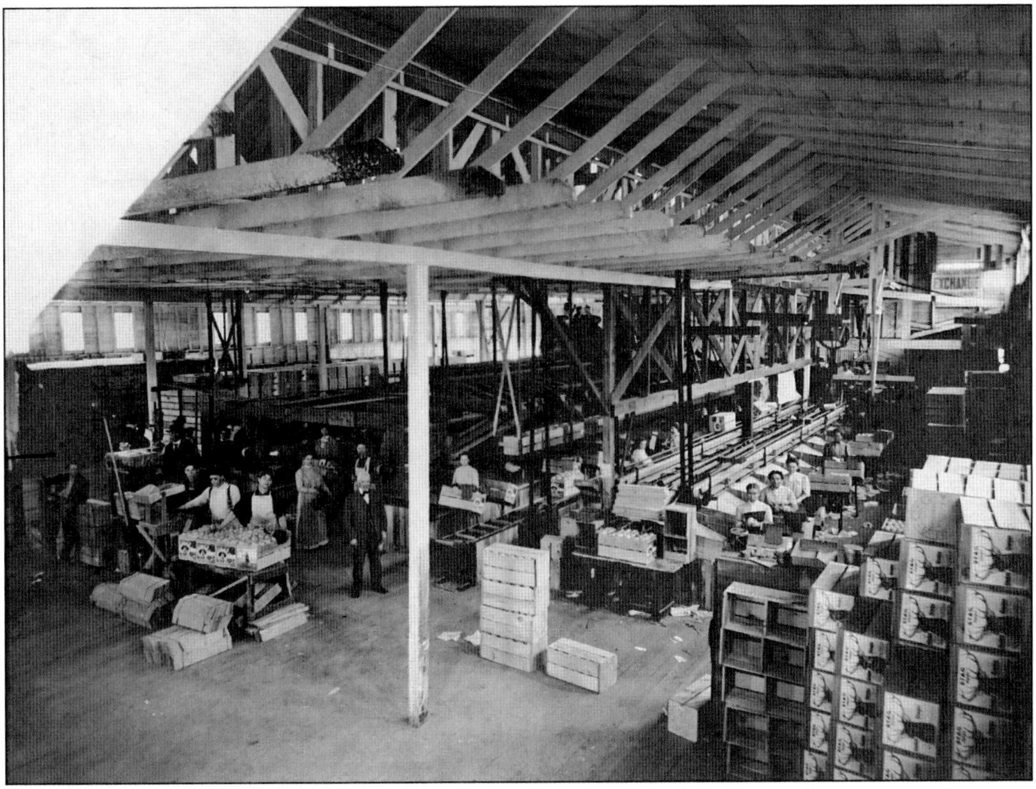

This interior photograph of the Glendora Citrus Association packinghouse shows crates of oranges marked with the Stag labels waiting to be shipped. (Courtesy of Glendora Historical Society.)

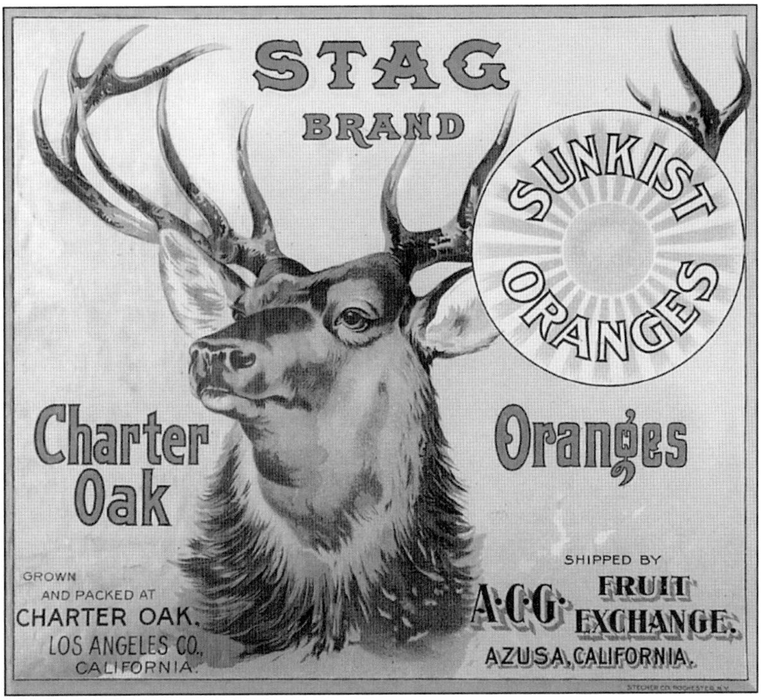

This is a later example of a Stag Brand label. The Sunkist medallion and the name Charter Oak have been added, as the oranges were also grown and packed in Charter Oak. The shipper was the ACG Fruit Exchange in Azusa. Lithograph companies kept unprinted versions of popular designs in stock so they could be ordered by different packinghouses. The Stecher Company in Rochelle, New York, printed the early Stag label.

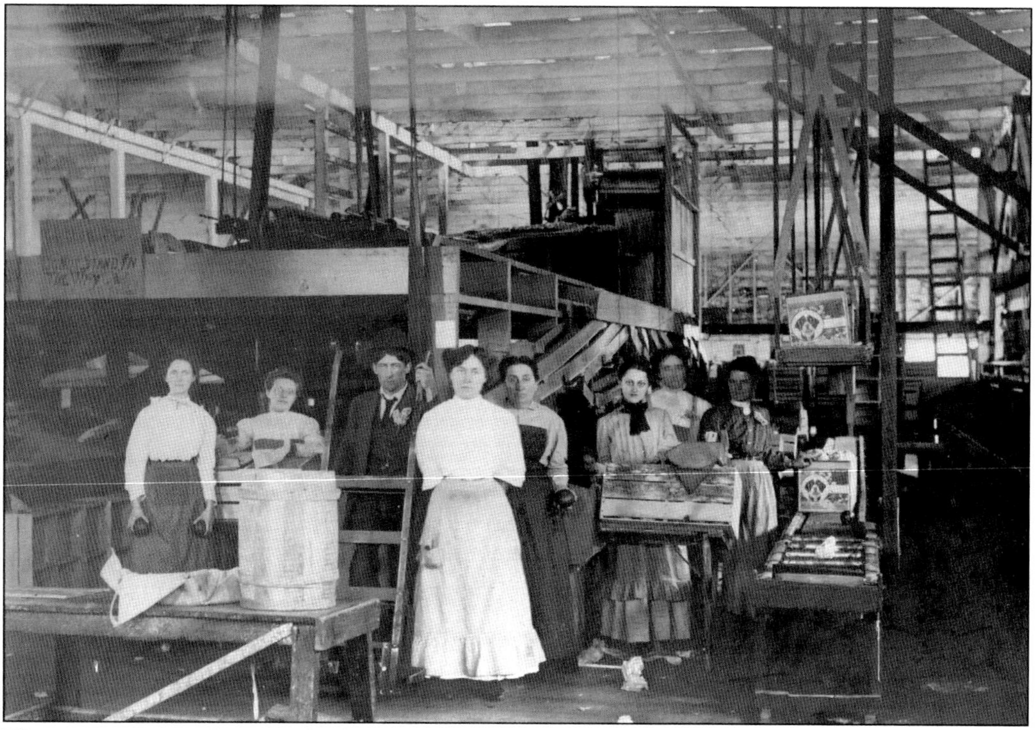

This is an interior photograph of a very early Glendora packinghouse. Only the lady packers and the manager (third from the left) are shown. The Pointer label is visible on the packing crate on the right side. (Courtesy of Glendora Historical Society.)

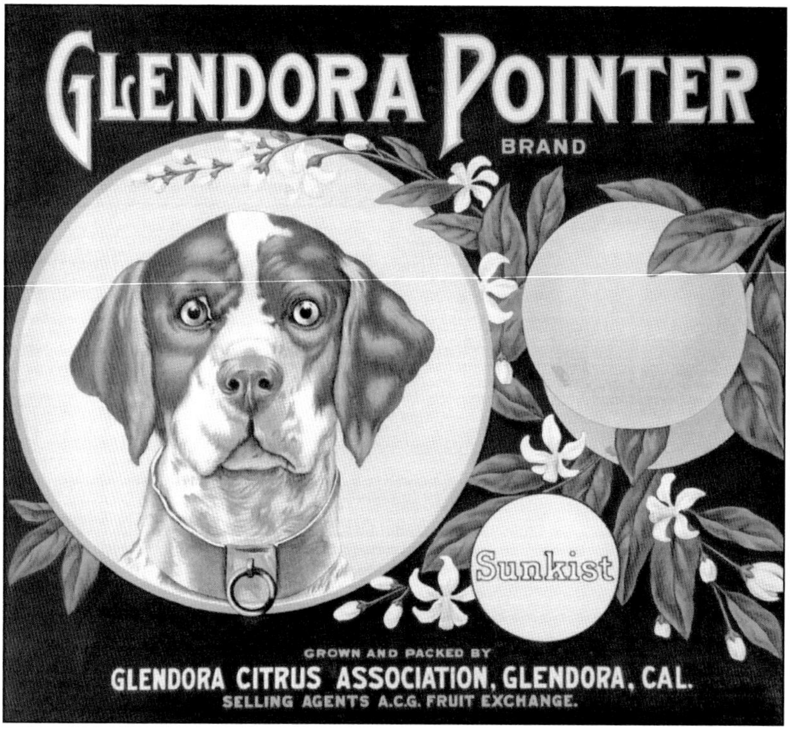

This is a late Glendora Pointer Brand label with the Sunkist medallion. ACG Fruit Exchange was the shipping agent. On March 27, 1905, the Southern California Fruit Exchange became the California Fruit Exchange. New members joined from the San Joaquin Valley. The exchange now had 5,000 member growers.

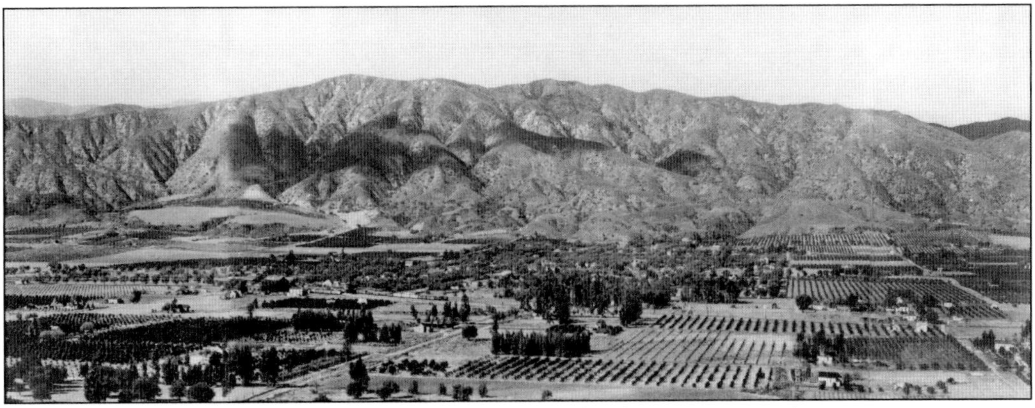

This photograph shows new, mature citrus groves in Glendora. Citrus production was increasing, and new markets were urgently needed to absorb the growth. E.O. McCormick, president of the Southern Pacific Railroad, offered to match dollar for dollar any amount the exchange would spend on advertising. The advertising campaign launched in Iowa by Lord and Thomas, a Los Angeles advertising agency, was the birth of Sunkist.

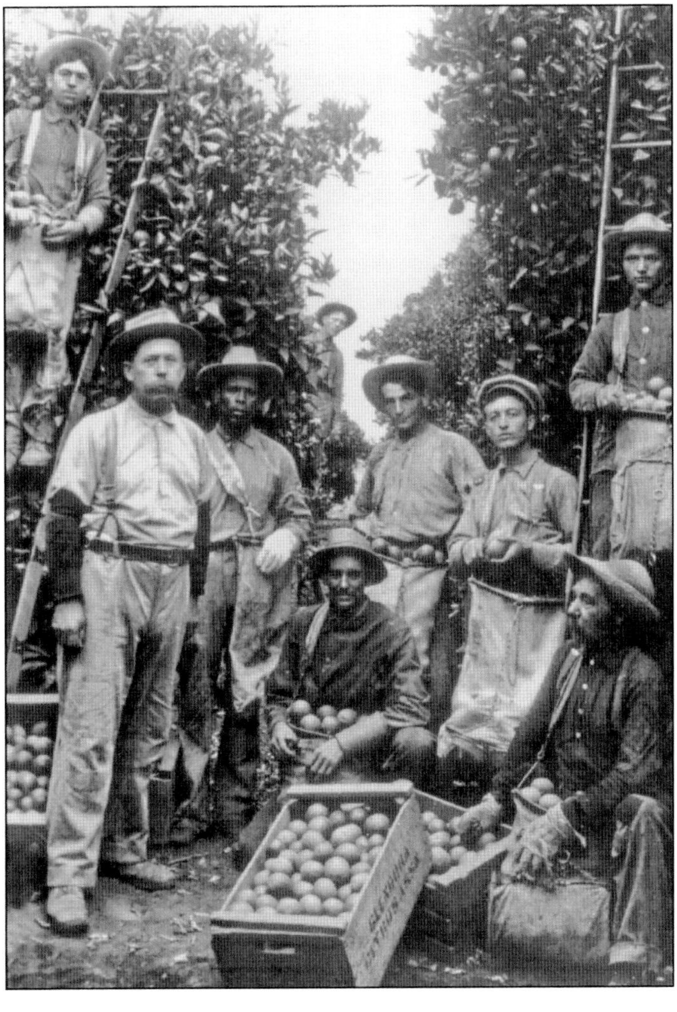

Glendora pickers are posed with their manager (standing second from left). The exchange advertising campaign increased citrus sales by 50 percent in Iowa. Citrus traveled from the Covina station in special bannered trains, whose arrival time at various Iowa stations was telegraphed ahead. The local advertisements for Iowa Orange Week appeared in the newspapers. The slogan, "Oranges for Health—California for Wealth," appeared on billboards. (Courtesy of Glendora Historical Society.)

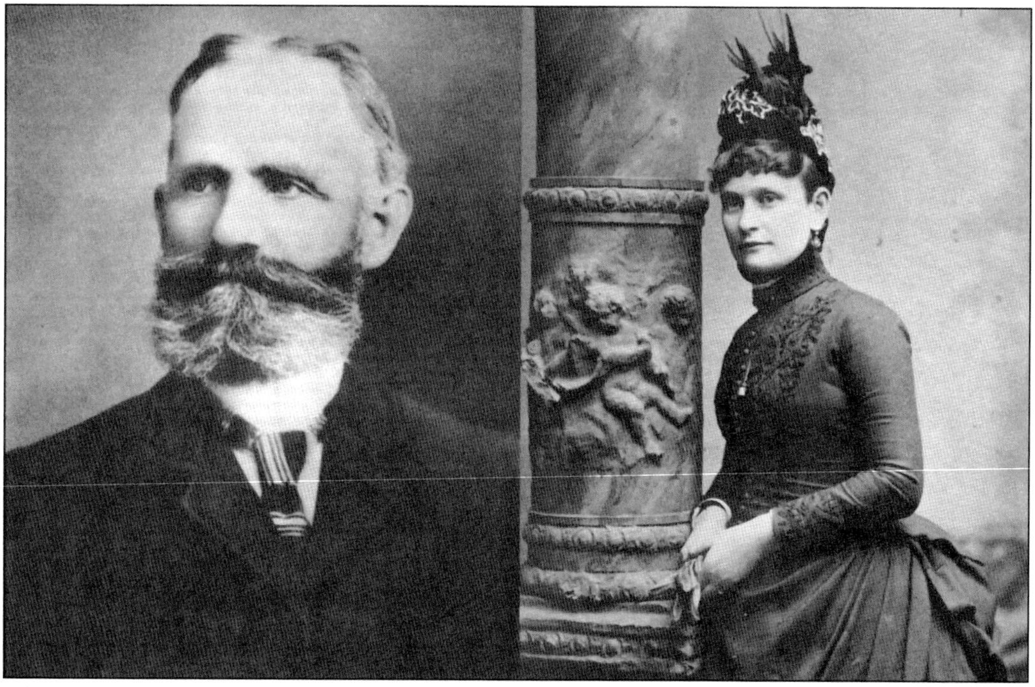

Alfred Patterson Griffith, photographed here with his wife, Alice Louisa Black, moved to California in 1891. He pioneered citrus growing in Azusa. Griffith began with 30 acres and acquired 200 more east of Azusa, where he built his own orange and lemon packinghouse, the A.P. Griffith Association. He was a member of the California Fruit Growers Exchange. He also pioneered Azusa water development as president of the Glendora-Azusa Water Company.

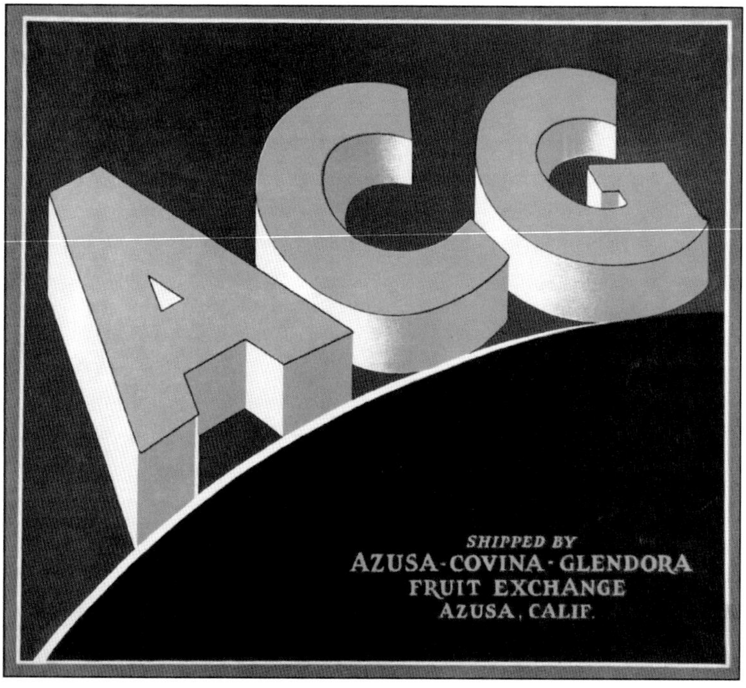

The first ACG packinghouses in Azusa and Glendora shipped under ACG labels for a while, but the ranchers preferred labels with the name of the packinghouse—Azusa, Covina, Glendora, Irwindale, or Charter Oak. The ACG became an exchange member of the California Fruit Growers Exchange.

Green Crusader Brand oranges were packed by the Azusa Citrus Association and shipped by the ACG Fruit Exchange, which became the shipping exchange for the entire valley when ranchers withdrew from the San Antonio Exchange.

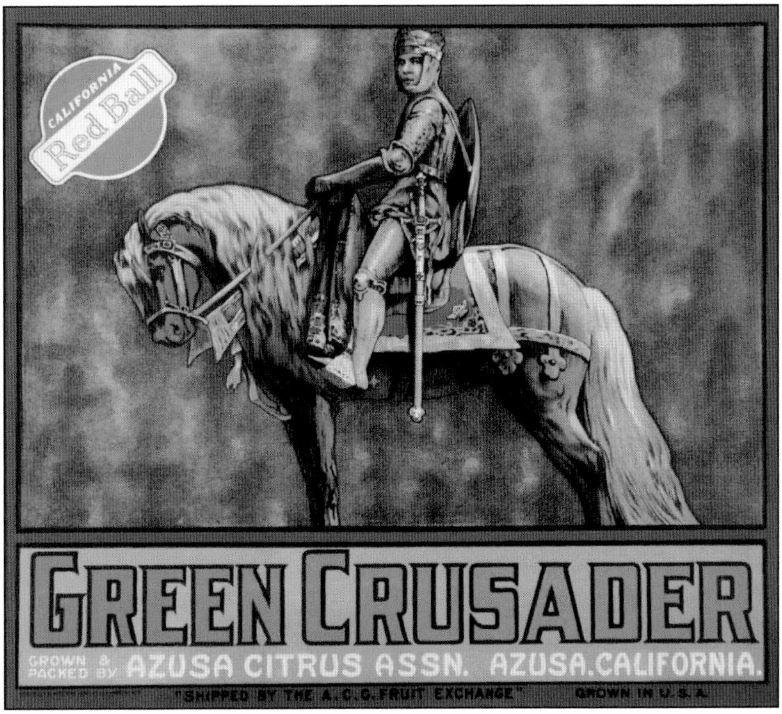

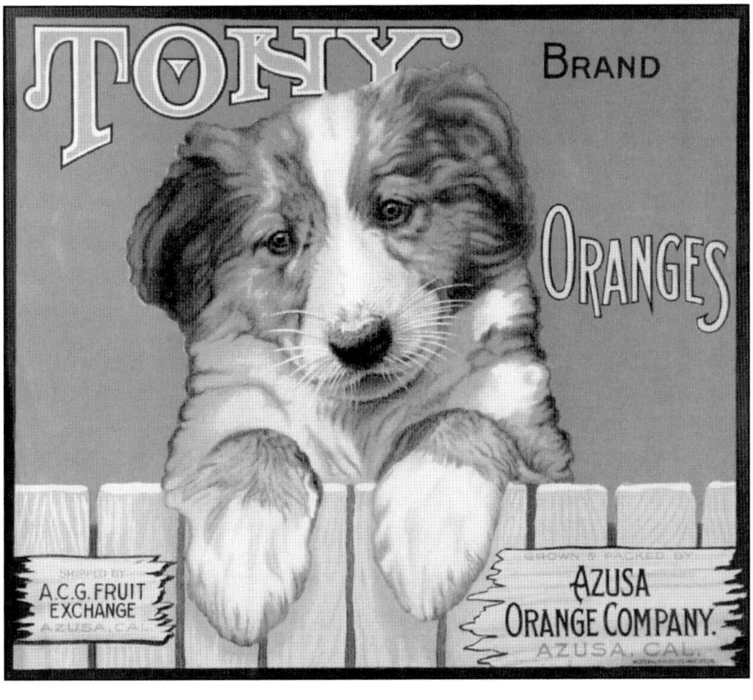

Tony Brand oranges were packed by the Azusa Orange Company. The label was printed by Los Angeles Western Lithograph Company. Azusa had five major packinghouses: the Azusa-Foothill Citrus Company, Azusa Citrus Association, Golden Orange Grove, Inc., Azusa Orange Company, and the A.P. Griffith Association. The Azusa Orange Company shipped through Mutual Orange Distributors, and the others shipped through ACG.

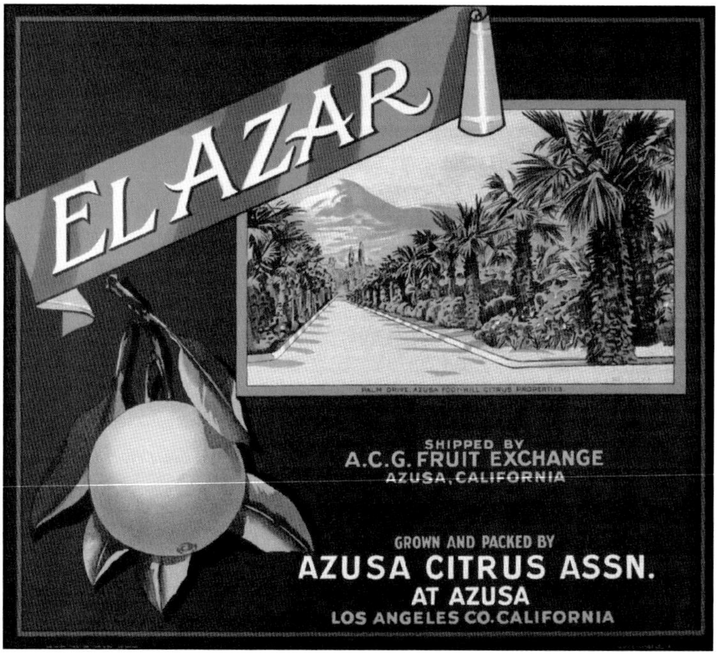

The El Azar label illustrates Azusa's famous double-palm drive leading to the homes of Kate Slauson Vosburg and Louisa Slauson Macneil. The drive was created because the feuding Slauson sisters did not want to see each other. The El Azar label is an interesting example of how citrus crate labels were not always accurate. In Spanish, the word *azahar* means orange or lemon blossom, but the word *azar* means chance or misfortune. Schmidt Lithograph printed the label for the Azusa Citrus Growers.

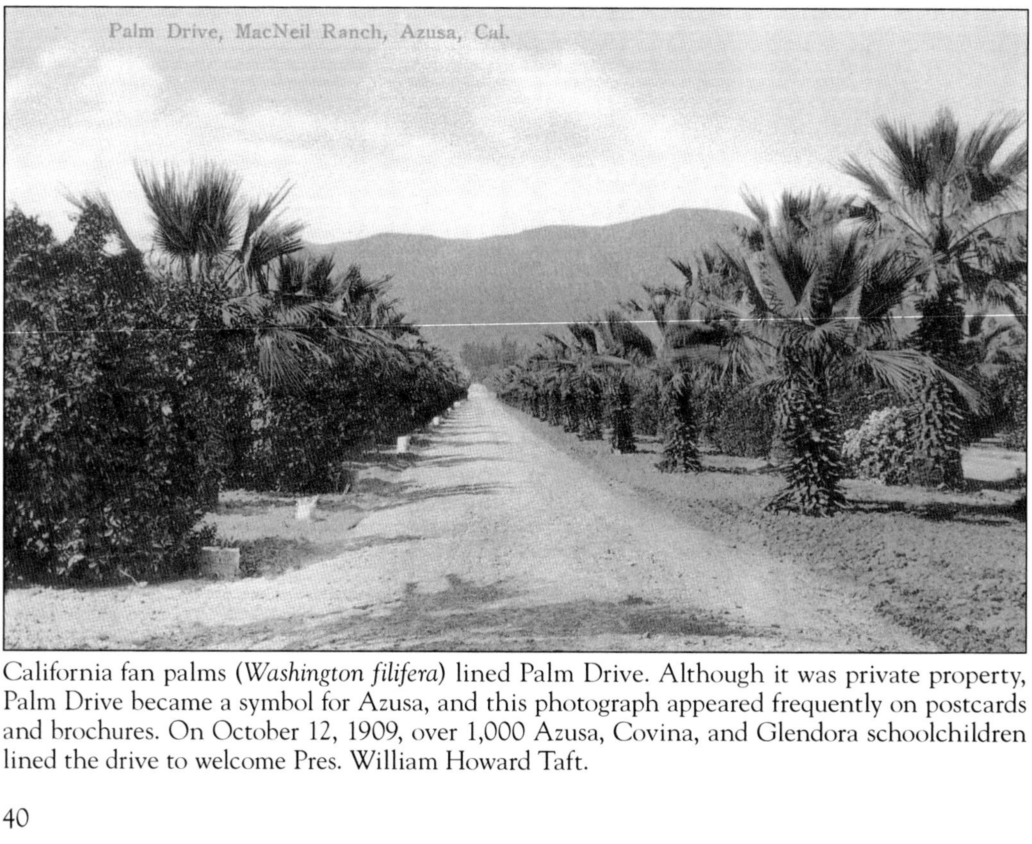

California fan palms (*Washington filifera*) lined Palm Drive. Although it was private property, Palm Drive became a symbol for Azusa, and this photograph appeared frequently on postcards and brochures. On October 12, 1909, over 1,000 Azusa, Covina, and Glendora schoolchildren lined the drive to welcome Pres. William Howard Taft.

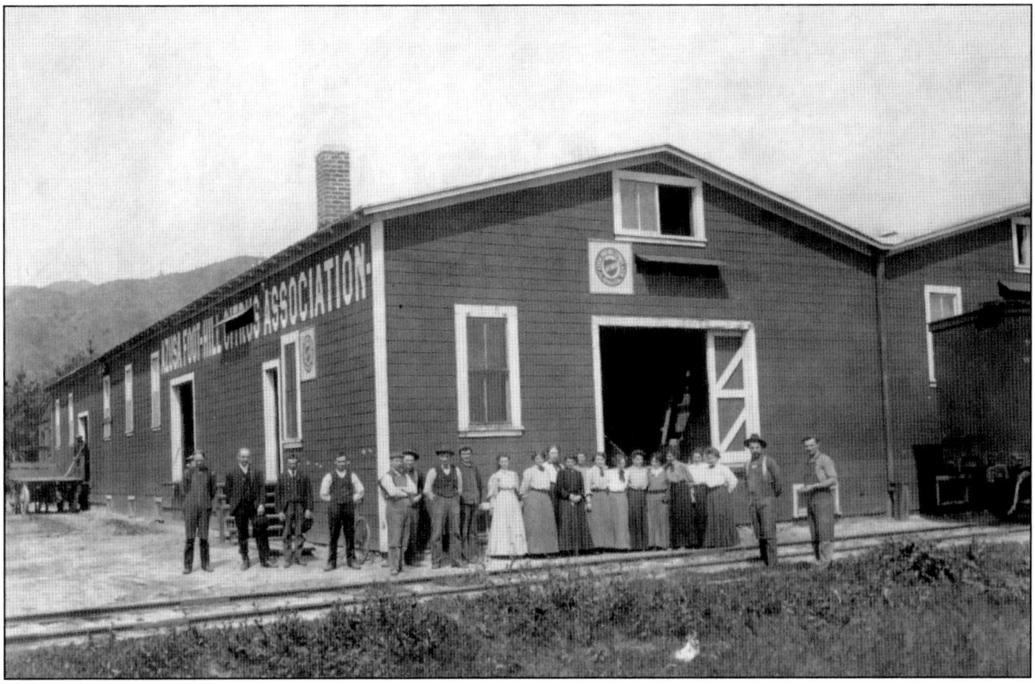

Jonathan Sayre Slauson's three children, Kate, Louise, and James, combined their citrus properties and created the Azusa-Foothill Citrus Company. This packinghouse north of the Santa Fe tracks between Soldano and Pasadena Avenues processed the fruit from the family's approximately 800 acres. Four buildings had loading docks, and inside was the latest packing equipment. (Courtesy of Jeffrey Lawrence Cornejo Jr.)

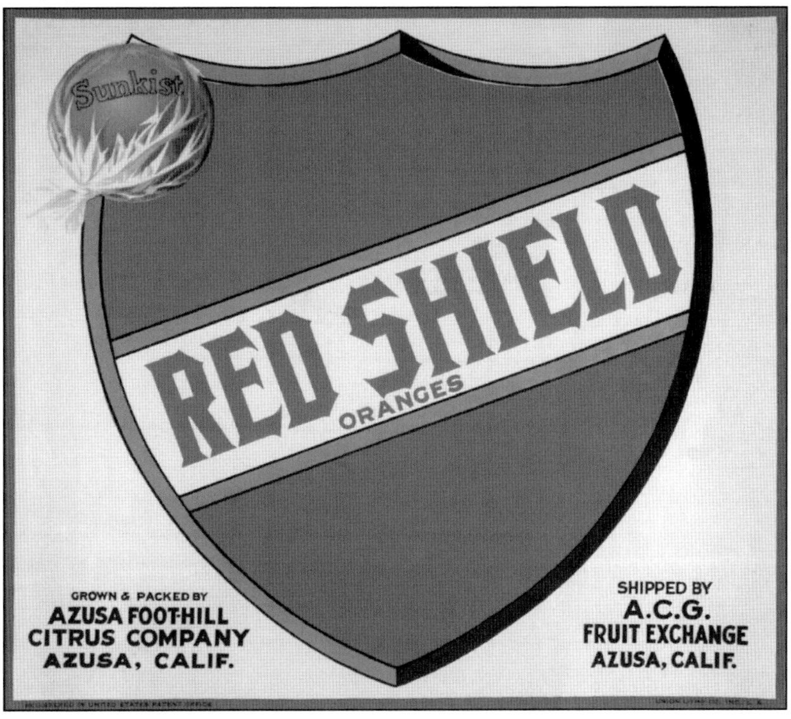

This is the Azusa-Foothill Citrus Company's Red Shield label, printed by the Union Lithograph Company. It has a tissue-wrapped, Sunkist-labeled orange with paper twisted around it.

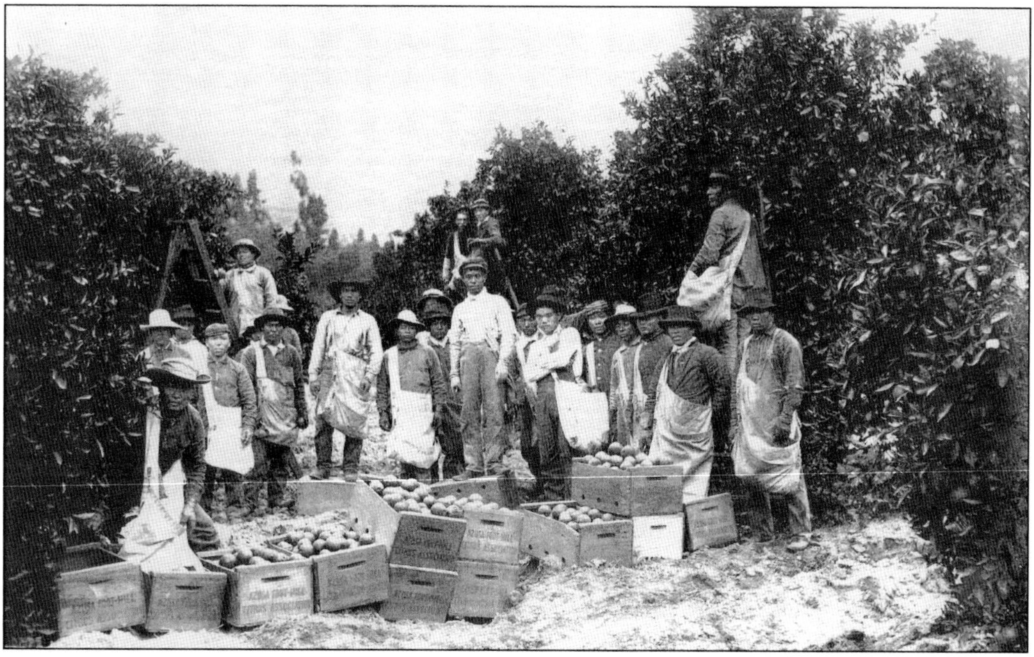

These pickers worked for the Azusa-Foothill Citrus Company. They were provided with lodging and meals and were paid for each field box they picked. The crew boss marked each field box with the initials of the pickers. (Courtesy of Azusa Historical Society.)

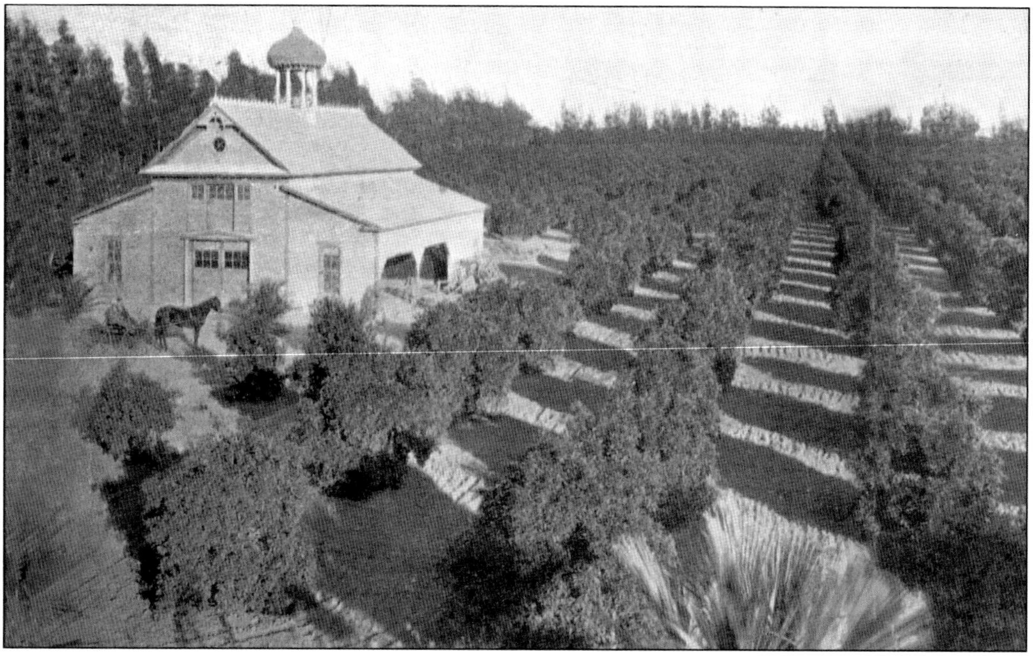

J.H. Adams's El Dorado Ranch, which began in 1886, was the first citrus grove planted on the Phillips Covina tract. Adams planted 40 acres of oranges and 5 acres of lemons. A tall hedge of eucalyptus and cypress trees surrounded the property. Outside the hedge were alternating pepper trees and palms. Beautiful tropical and semitropical plants also surrounded the house. Adams marketed his fruit through the Earle Fruit Company.

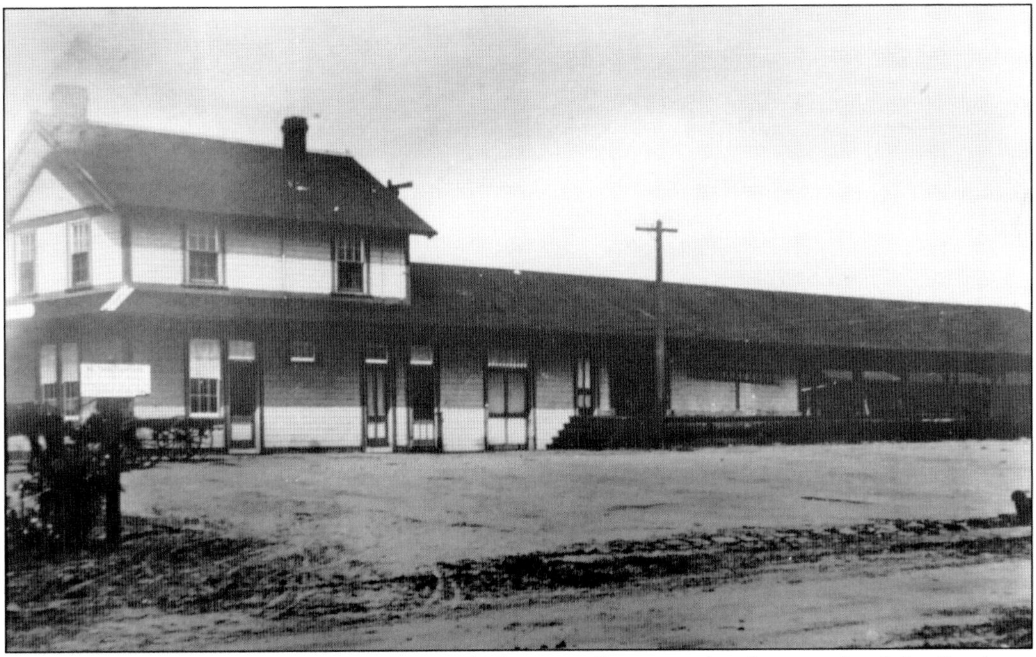

On August 15, 1895, two crews of workers and 75 teams of horses and mules completed the Southern Pacific tracks and telegraph line to Covina. Work on the tracks began in 1894. A boxcar served as a ticket office until this two-story depot was built. The stationmaster and his family lived on the second floor. When passenger service started, a round-trip ticket to Los Angeles cost $1.15.

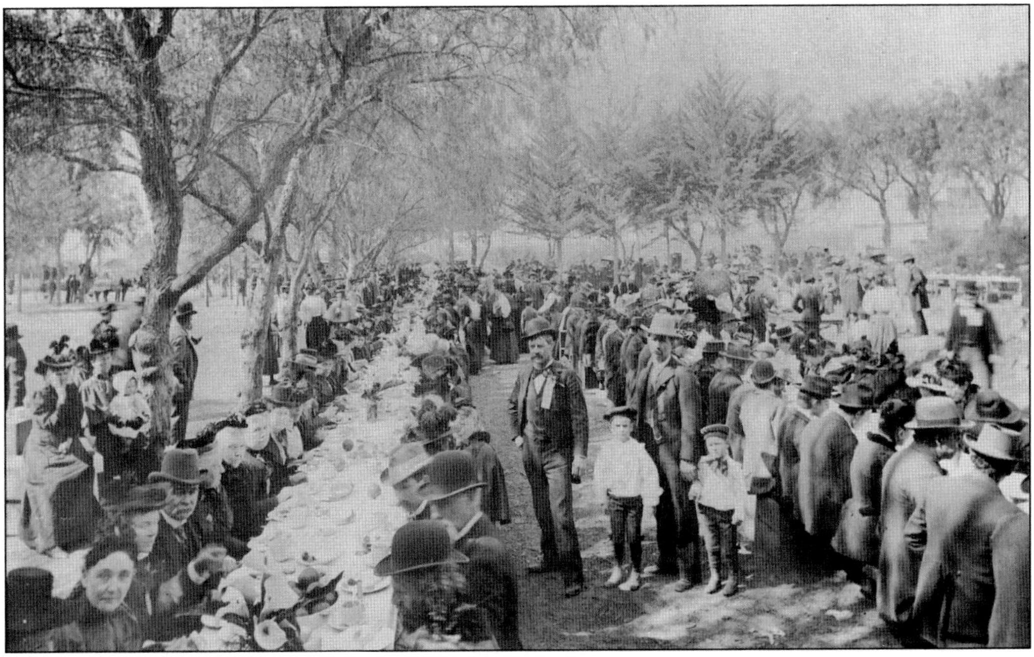

Soon after the first train came to Covina, hundreds of families gathered for a barbeque to celebrate the Southern Pacific Railroad's arrival. Tables were decorated with calla lilies and displays of fruit. Ladies wore their best hats. Speeches were given praising those who had worked so hard to bring the railroad to Covina. A concert and a baseball game ended the day.

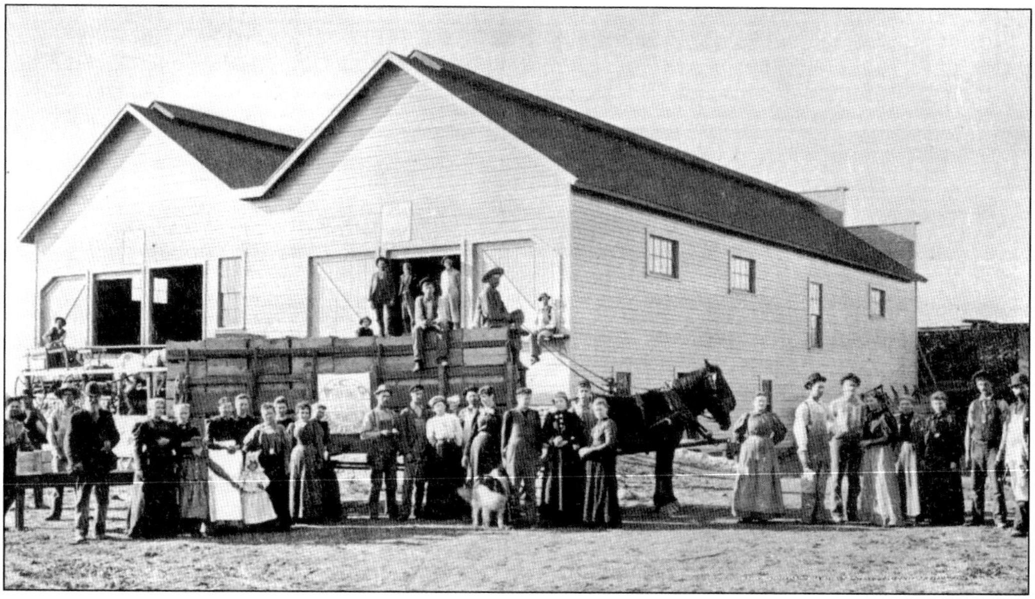

On September 9, 1895, a special 12-car train pulled by a Southern Pacific engine left Covina for the port of Los Angeles beach (Santa Monica) for a day of bathing and picnicking. The train was decorated with chains of flowers and banners that read, "Welcome SPRR and Covina: Garden Spot of California." The Covina Citrus Association packinghouse was the first built after the railroad arrived.

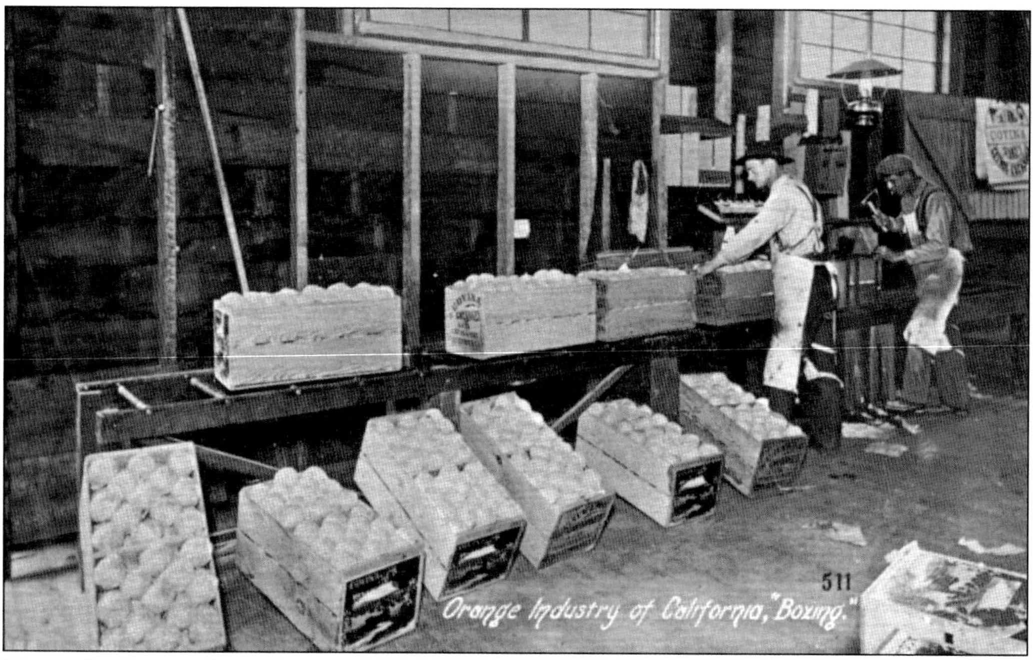

This is the Covina Citrus Association basement where the shipping crates were made and labeled. The men are preparing crates to load on a boxcar. In the early years of the Southern California Fruit Exchange, some ranchers who had joined the exchange broke their contracts. There was no way to enforce contracts because the exchange was not incorporated. On October 16, 1895, the exchange directors authorized the incorporation.

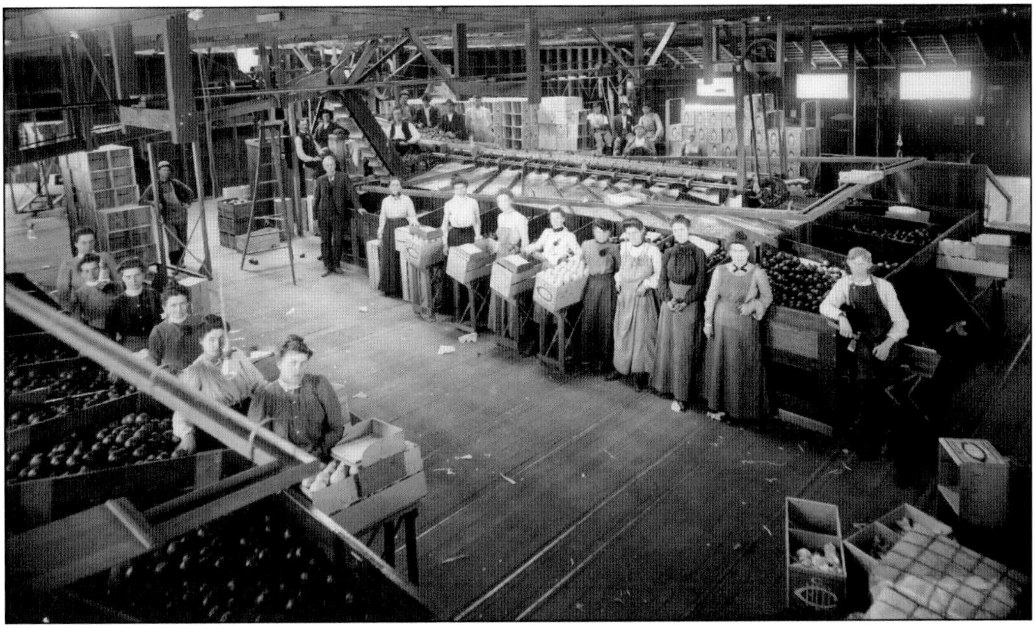

The Covina Citrus Association was incorporated in the fall of 1895. They built this two-story packinghouse beside the Southern Pacific Railroad tracks in Covina. The first board of directors included president Thomas Francis Griswold, secretary H.E. Cheseboro, foreman E.G. Clapp, and treasurer Covina Bank. Directors included Lambert Ratekin, Anton C. Kerckhoff, R. Meredith, J.C. Prescott, S.A. Overholtzer, and George E. Anderson.

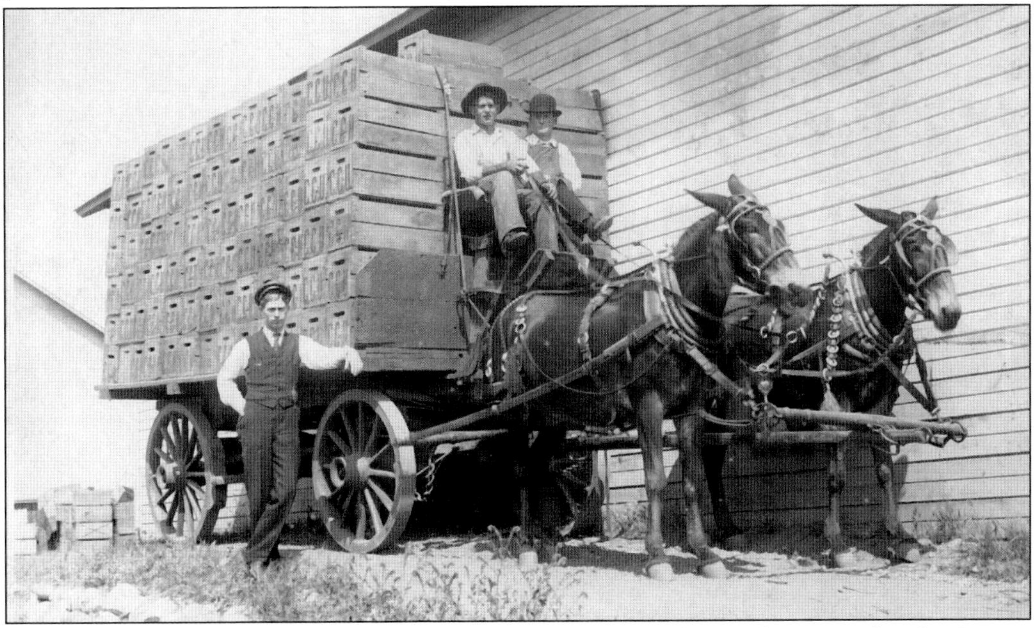

A wagon pulled by two mules and loaded with orange-filled field boxes has arrived at the Covina Citrus Association packinghouse. Standing beside the wagon is the packinghouse clerk who will check in the shipment and give the grower credit. The boxes were transferred from the wagon to a conveyor belt to be processed or stored. Growers returned to the packinghouse to pick up the empty boxes.

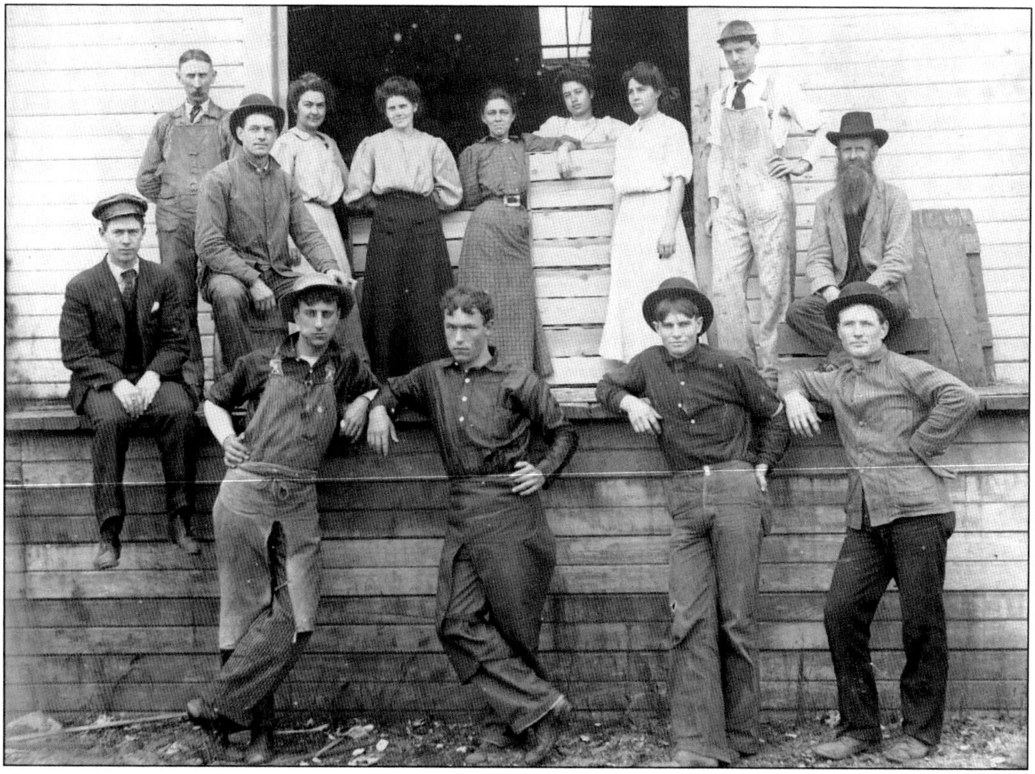

Covina Citrus Association employees are posed outside the packinghouse. William Hoogendyk wrote in *Women as Packers in Orange Houses*, "Many wives and daughters of the pioneers helped pack oranges. We honor them for the necessary part they took in making the barren desert of yesterday the garden spot of today."

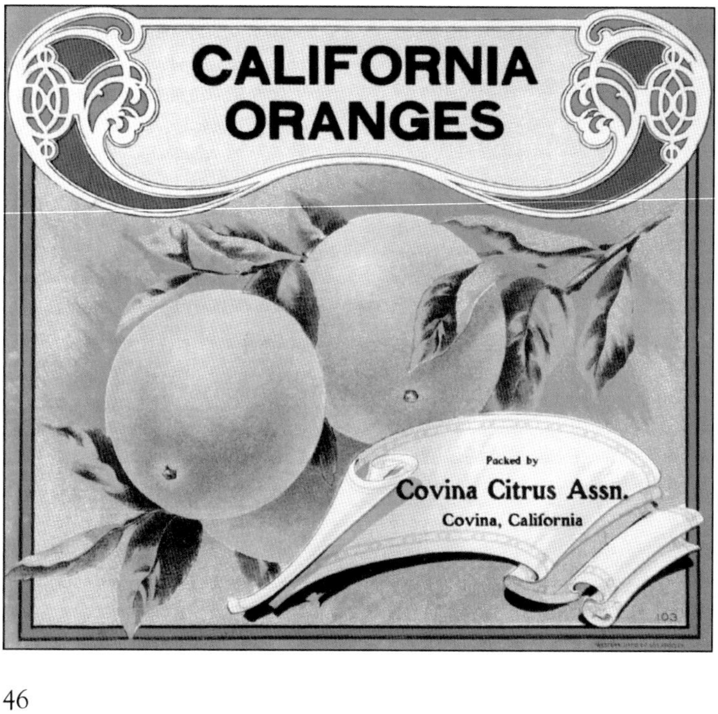

The California Oranges label was printed by Los Angeles Western Lithograph Company for the Covina Citrus Association. This early label does not have a wrapped Sunkist orange or a medallion because the trademark had not been created yet.

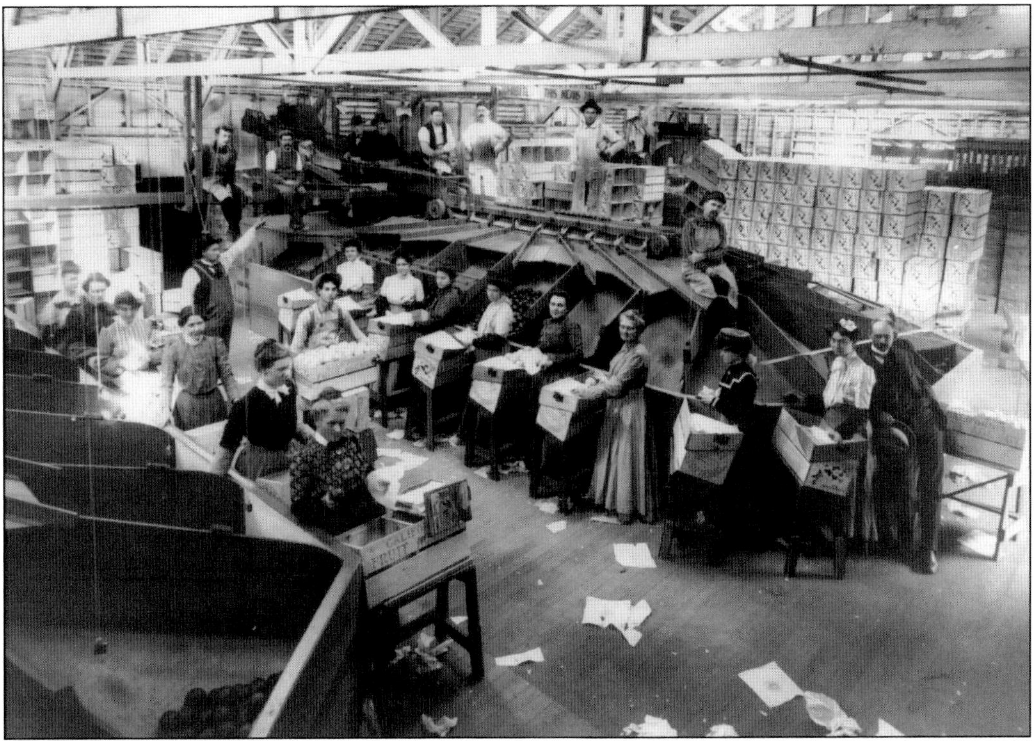

This photograph shows the first Covina Orange Growers Association packinghouse. The association packed under the Daffodil, Orchard King, Daisy, Gladiola, Preferred, Nonpareil, Par Value, Salvia, and Holly labels.

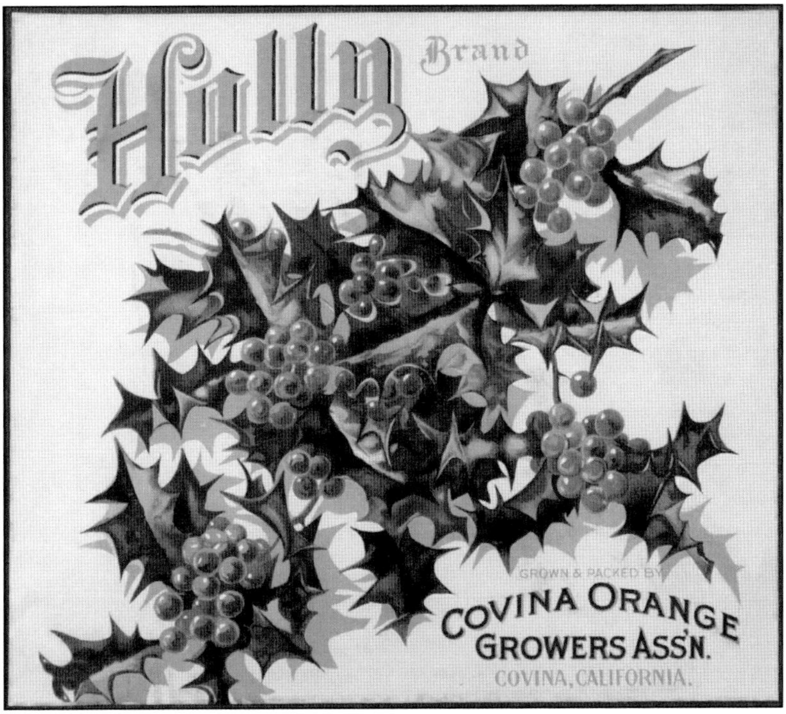

The Holly Brand label was designed for the Covina Orange Growers Association. It was the second cooperative citrus association formed in Covina. In 1908, the California Fruit Exchange Board of Directors ordered six million stickers for Sunkist oranges and one million stickers for Sunkist lemons to be printed and posted on the regular labels of exchange shippers.

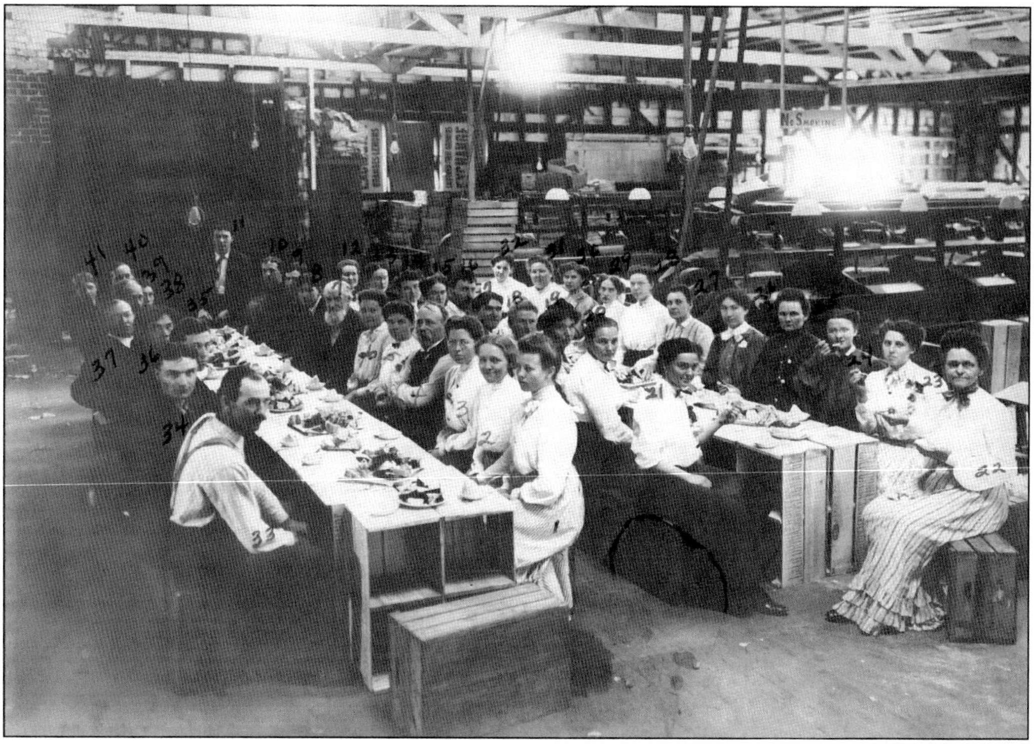

The packing crew has gathered for a celebration at the Covina Orange Growers Association packinghouse. Manager Thomas Finch is seated fourth from the left in the second row.

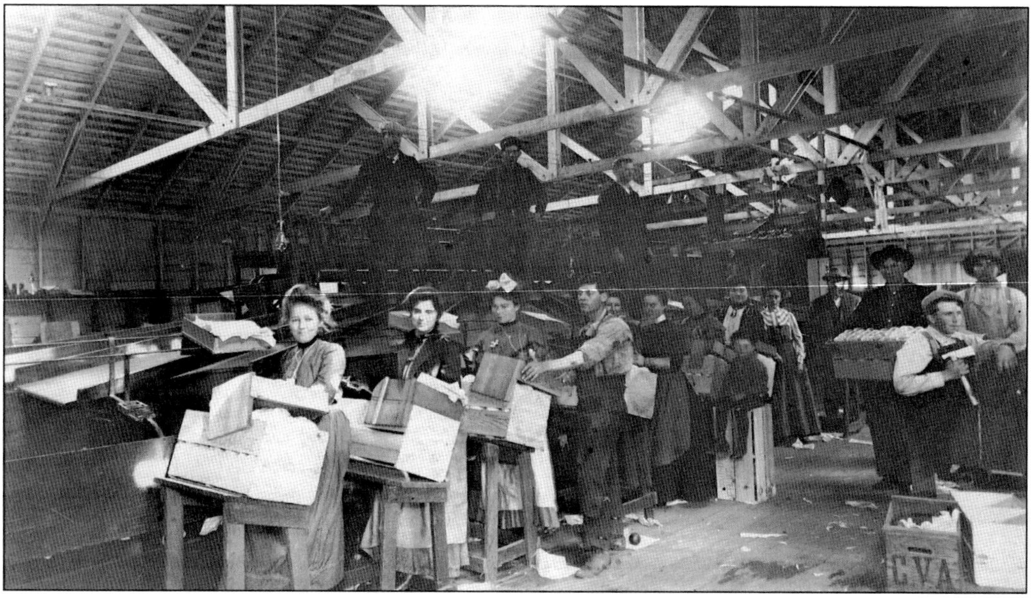

Inside the Covina Orange Growers packinghouse, oranges started along the conveyor belt to be washed and blown dry. Culls were then separated into bins to become citrus by-products like pectin and fertilizer. Rows of women wrapped good fruit in tissue paper and placed it in packing boxes. The lids were nailed shut, and the boxes were refrigerated for shipping. Thomas Finch and Lambert Ratekin were the managers.

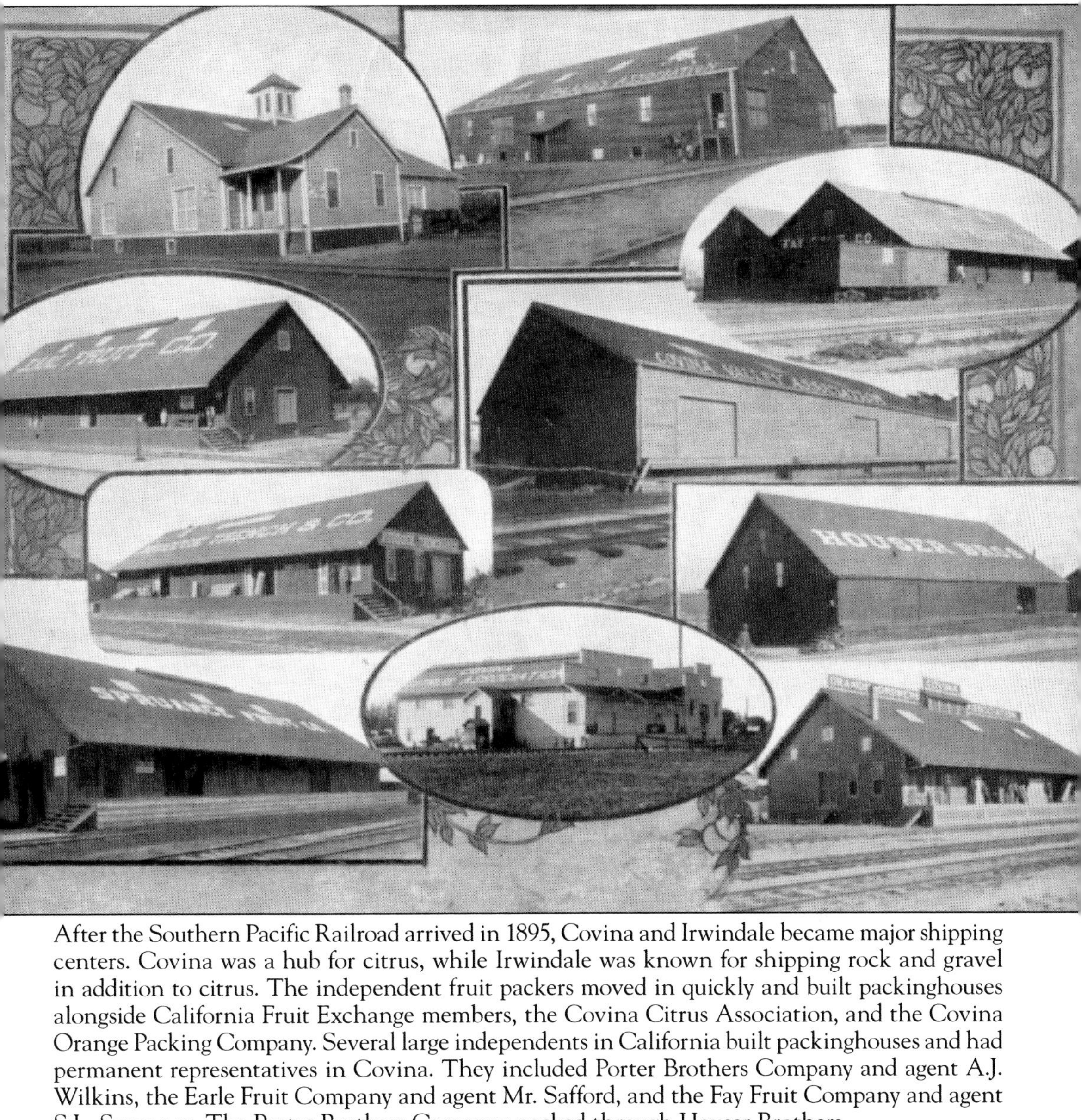

After the Southern Pacific Railroad arrived in 1895, Covina and Irwindale became major shipping centers. Covina was a hub for citrus, while Irwindale was known for shipping rock and gravel in addition to citrus. The independent fruit packers moved in quickly and built packinghouses alongside California Fruit Exchange members, the Covina Citrus Association, and the Covina Orange Packing Company. Several large independents in California built packinghouses and had permanent representatives in Covina. They included Porter Brothers Company and agent A.J. Wilkins, the Earle Fruit Company and agent Mr. Safford, and the Fay Fruit Company and agent S.L. Spruance. The Porter Brothers Company packed through Houser Brothers.

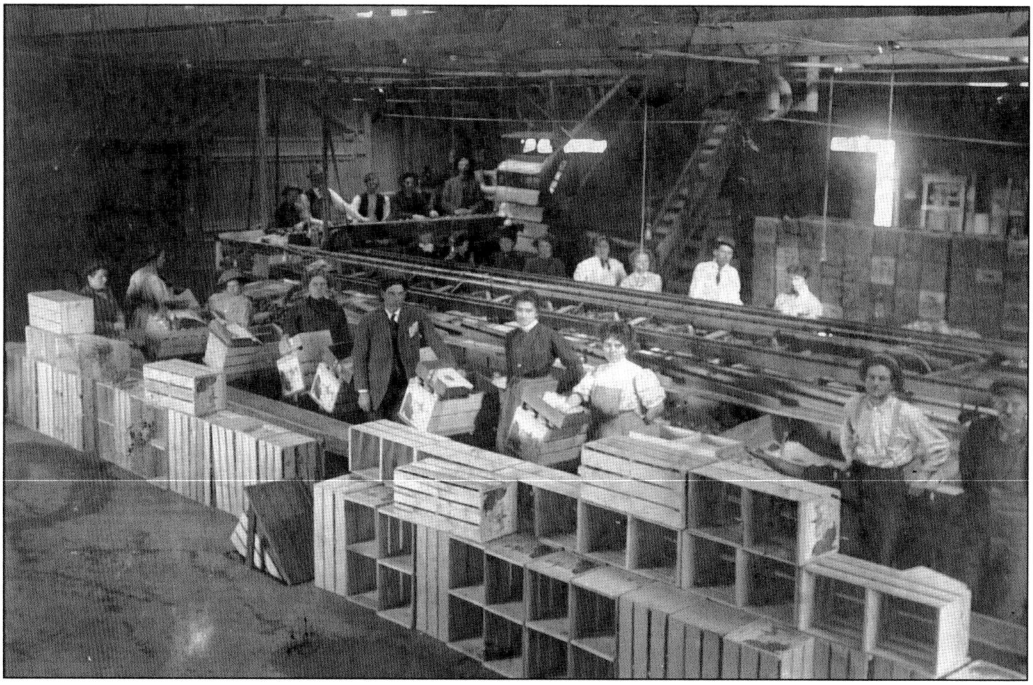

The Riley Fruit Packing Company was an independent packinghouse located on Azusa Avenue between San Bernardino Road and Cypress Street. It was purchased by Harry Damerel and later became affiliated with the California Fruit Exchange.

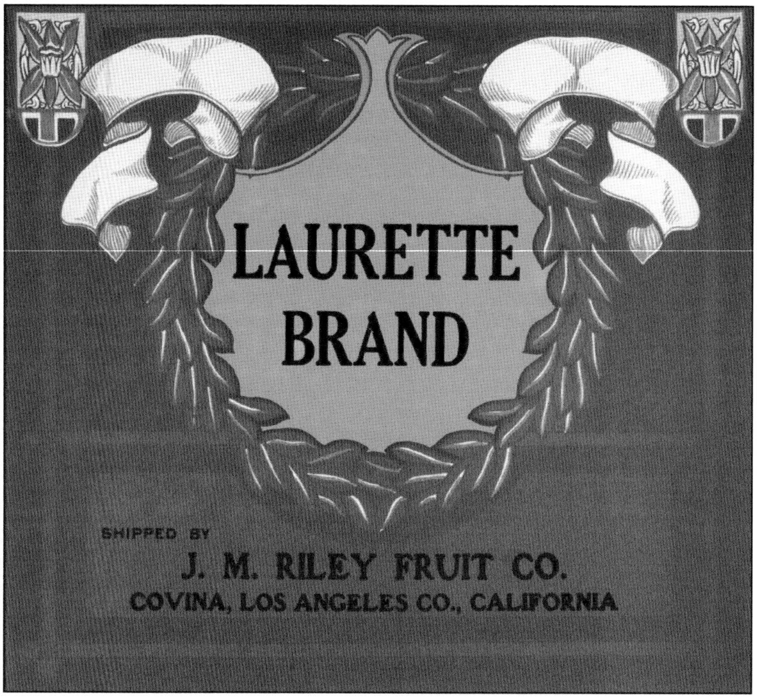

The Schmidt Lithograph Company in Los Angeles printed the Laurette label for the Riley Fruit Company.

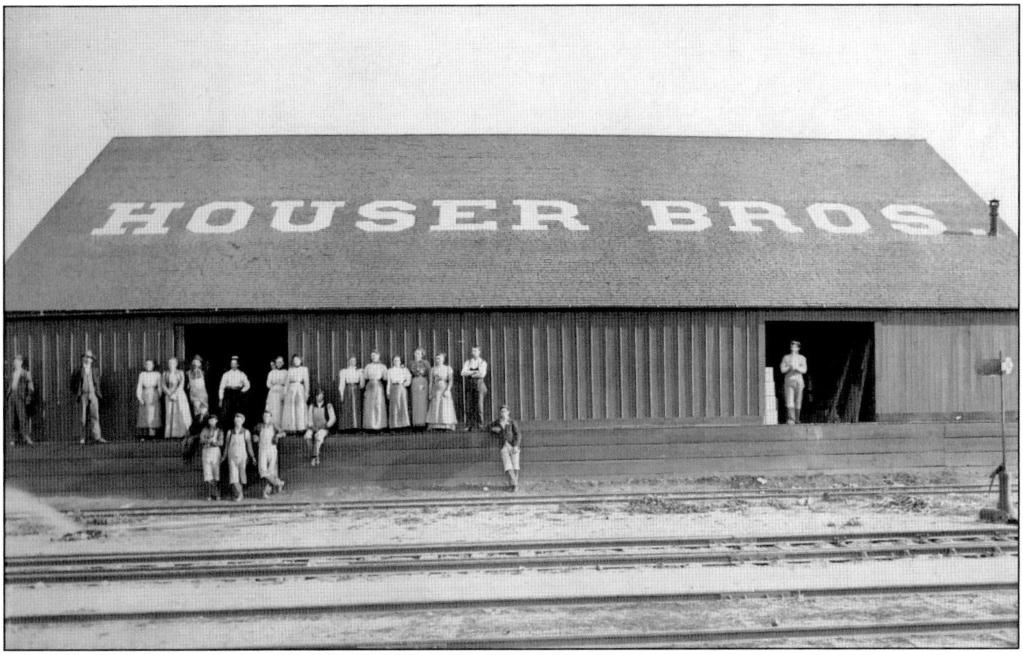

After the Southern Pacific Railroad arrived in Covina, the independent Houser Brothers packinghouse was located east of the train depot. The week after they opened, they were shipping a carload of oranges a day. The Houser family had a long history in Covina. Daniel Houser, who invented the combined harvester, helped finance the ditch that brought water to Covina.

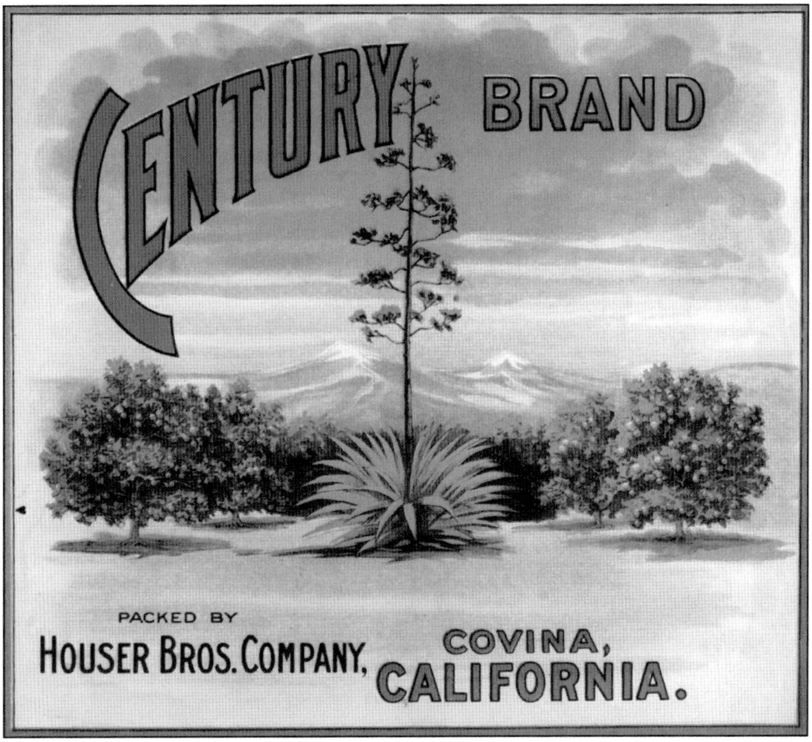

Century Brand is a rare Houser Brothers label. It illustrates the yucca plant that blooms every 100 years against a background of snow-covered hills and blooming orange trees. The Yucca leaves are sword-like as the designer has shown correctly, but the bell-shaped flowers on the stem are not shown properly. The yucca was also called the candle of the lord.

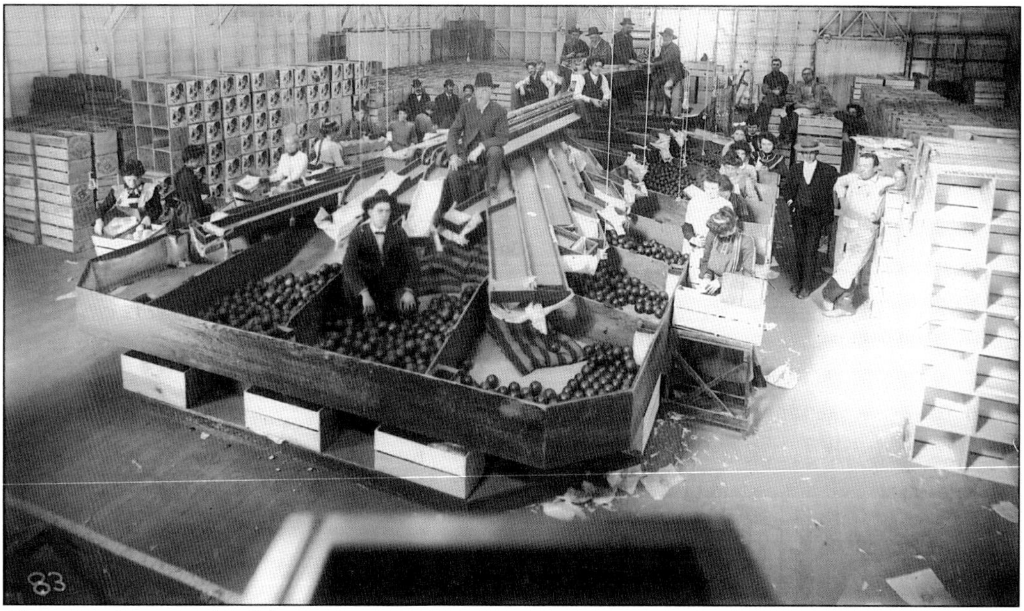

This is a photograph of the interior of the Houser Brothers packinghouse in Covina. Everyone has been carefully posed. John Houser, in suit, hat, and tie, is seated on top of the fruit distribution machine.

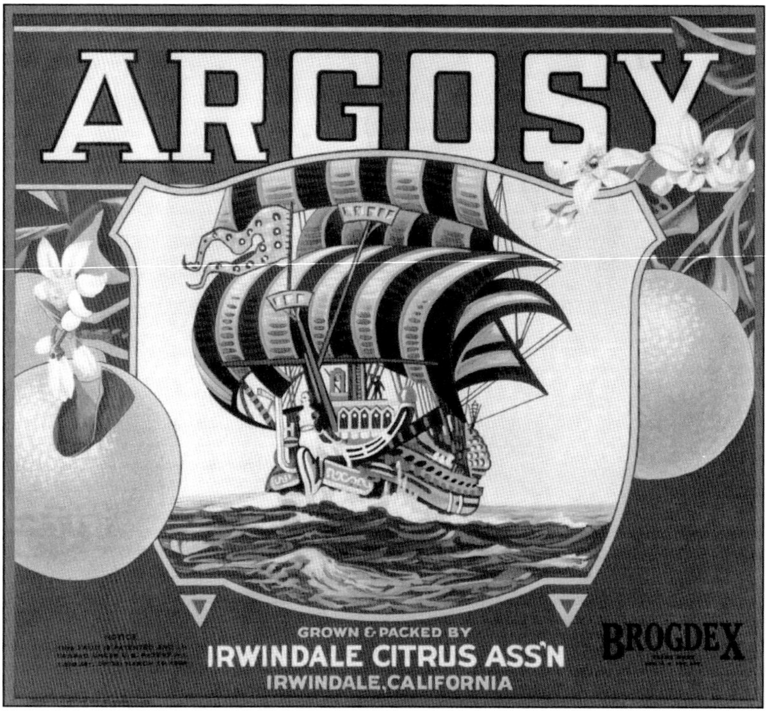

The Argosy label was printed for the independent Irwindale Citrus Association by Los Angeles Western Lithograph Company. They also printed the Sunny California label for them after they joined the ACG Fruit Exchange. Crocker-Union of Los Angeles printed their Highest label.

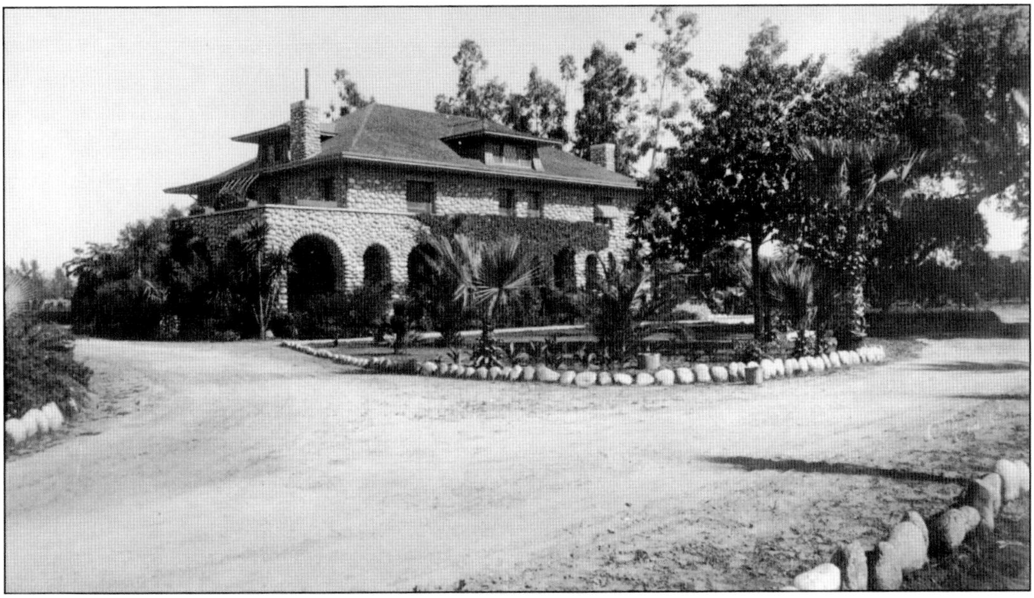

In 1894, Dr. Frank Marion Chapman moved his family to the Etna Green Ranch at Cypress Street and Grand Avenue. They built a three-story house of San Gabriel River rock with a ballroom on the third floor. The floors were oak with parquetry of cherry, maple, and ebony. Local landscapes were painted on the dining room ceiling. The ranch had its own packinghouse. In 1955, the house was torn down to make way for a subdivision.

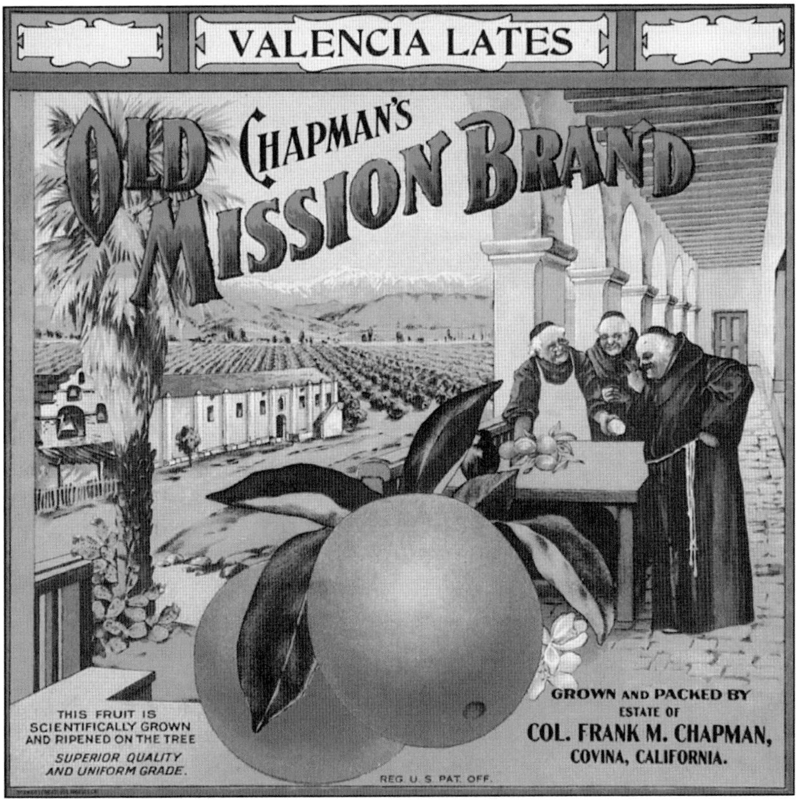

Chapman's Old Mission label was printed for the estate of Col. Frank Marion Chapman, who had his own packinghouse. It noted, "This fruit is scientifically grown and ripened on the tree; superior quality and uniform grade Valencia Lates."

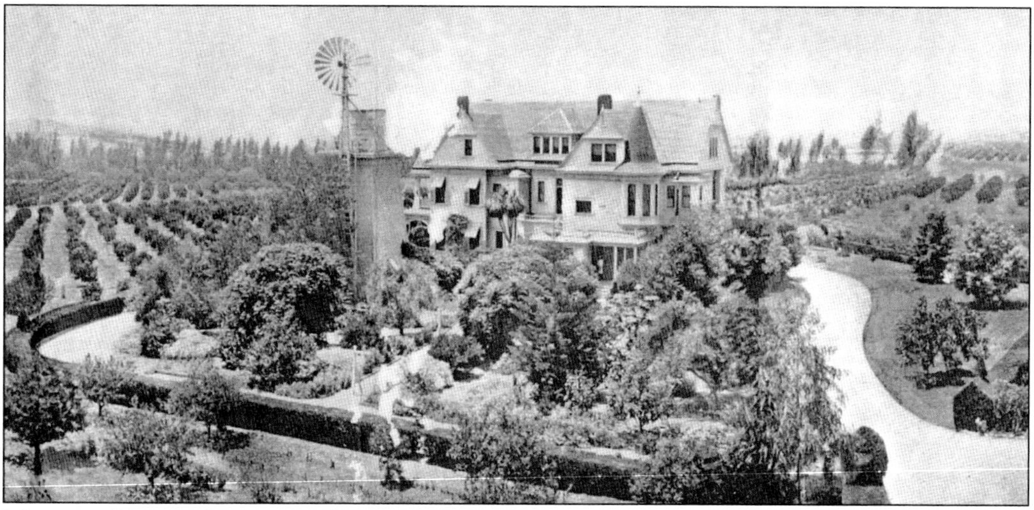

Mountain View, a three-story, Queen Anne–style mansion set in 400 acres of citrus and gardens, had 11 bedrooms, 5 bathrooms, and 7 fireplaces made of Belgian tile and rosewood. A stained-glass window looked down on the staircase, and the ballroom featured stained-glass chandeliers. Surrounding the mansion were stables, a carriage house, a bunkhouse, servants' quarters, and a caretaker's cottage. The 800-foot-long driveway was lined with palm trees and roses.

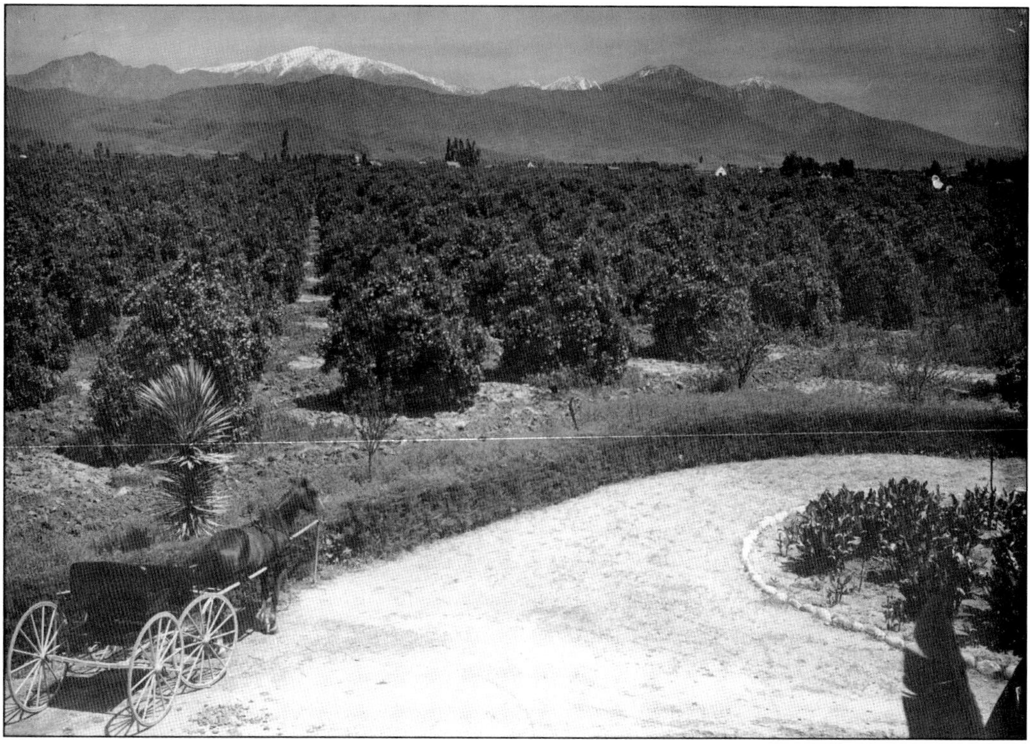

This photograph taken from the entry terrace looks toward the San Gabriel Mountains and the citrus groves of Glendora. Mountain View was the winter home of Charles H. Ruddock and his widowed mother. Her husband, Thomas S. Ruddock, selected the property in 1888. It was surrounded by 400 acres of citrus. The family also owned 400 acres of deciduous trees a mile and a half west of Covina.

This rare Mountain View label shows two oranges and two lemons with their blossoms against a background of the snow-covered San Gabriel Mountains with a road going up toward the top. It reads, "Mountain View grown and packed by the Ruddock Company, Covina." Charles Ruddock and his neighbor Frank Chapman had private packinghouses, and the Southern Pacific Railroad built a spur to reach them.

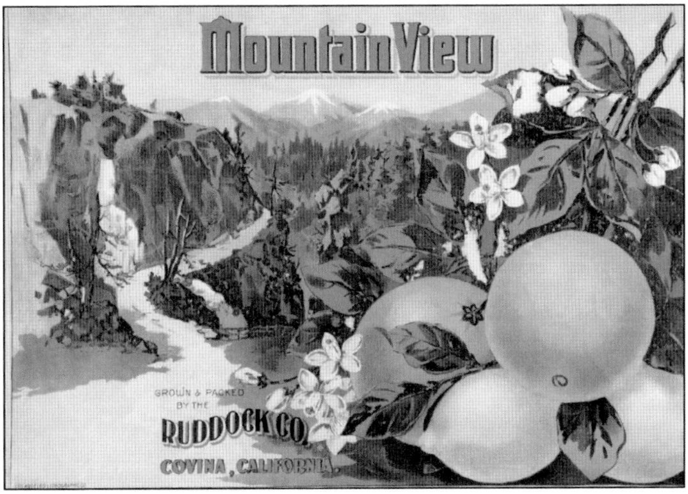

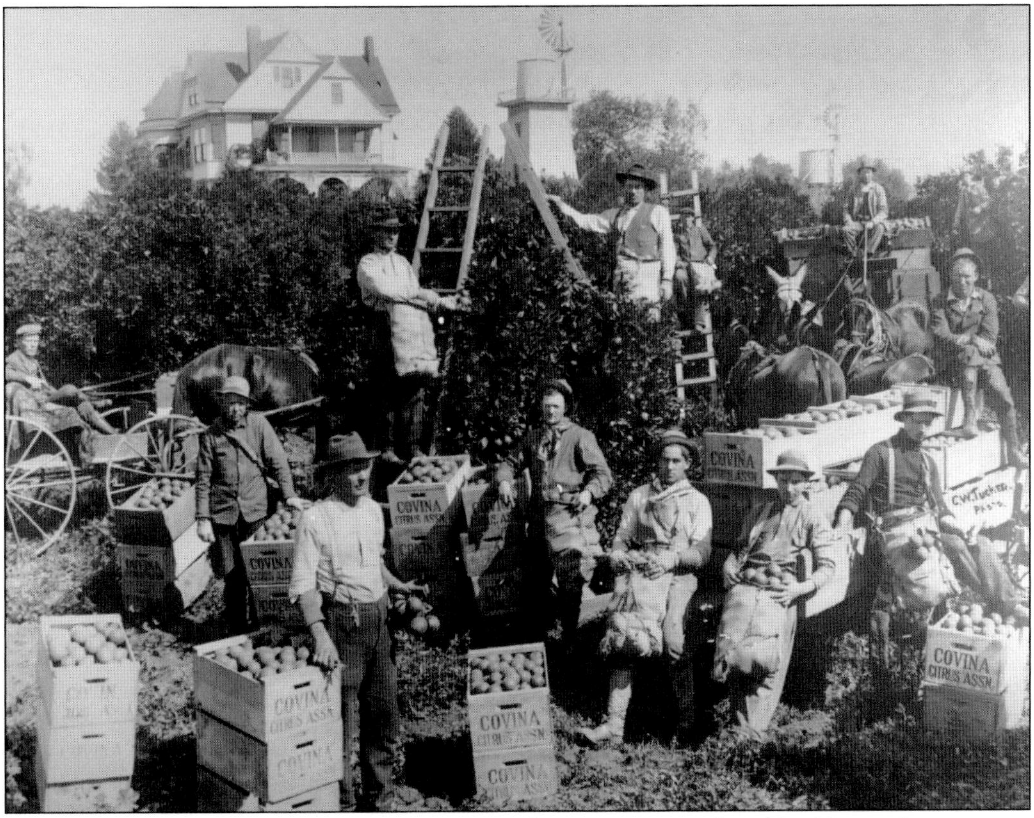

These men are picking oranges at Mountain View, the winter estate of wealthy Chicago businessman Charles Ruddock. After oranges were picked, they were loaded into field boxes on a wagon going to the packinghouse. These packers are using Covina Citrus Association field boxes, which meant Ruddock was not using his packinghouse for these oranges. The man in the horse-drawn buggy is either Charles Ruddock or Orville Vincent, the supervisor of the estate.

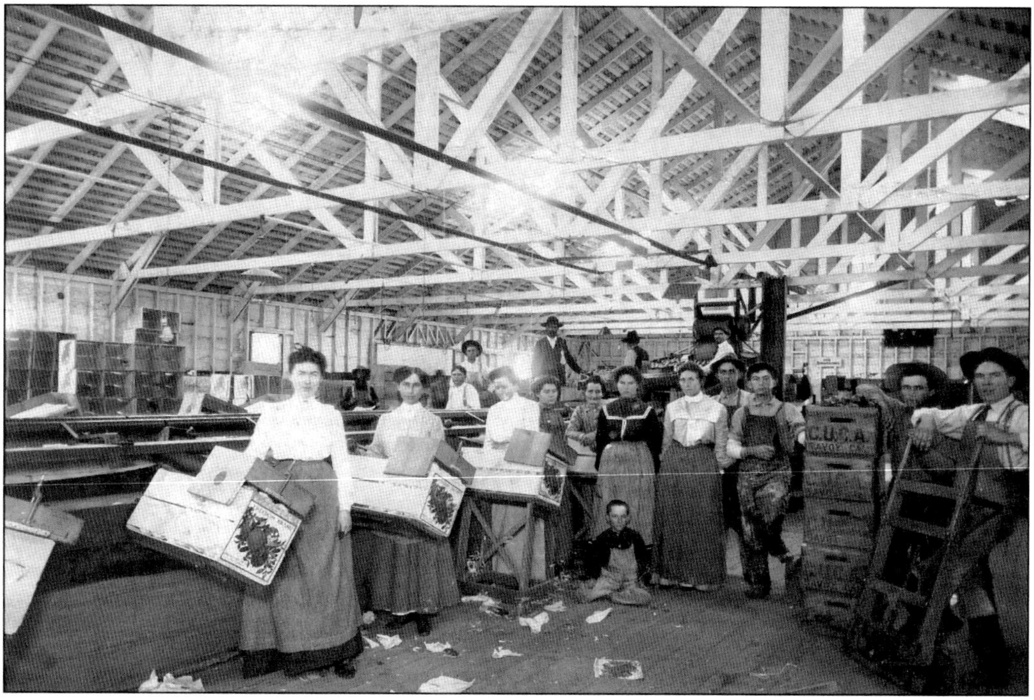

This is the interior of a very early packing operation, the Independent Fruit Company packinghouse at Savoy on the south side of the Southern Pacific Railroad tracks at Banna Avenue and Cypress Street. The foreman was J.L. Motter. Notice the Freedom label on the sides of the shipping crates. A young boy in work clothes is seated on the floor; he appears to be part of the crew.

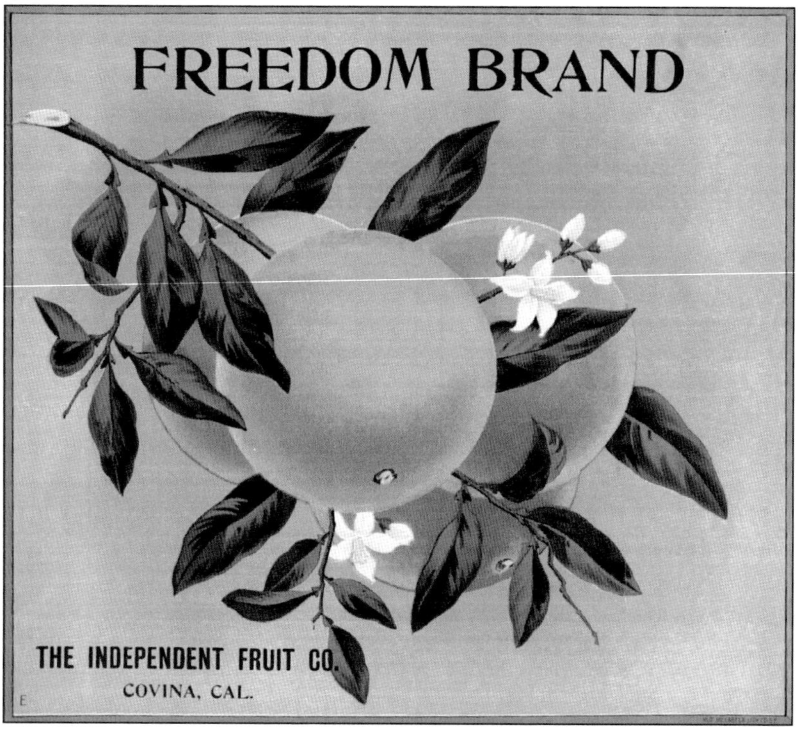

This rare Freedom Brand label shows ripe oranges with blossoms. "The Independent Fruit Company, Covina Cal," is all that is printed on the label. The majority of Covina Valley independent packinghouses were along the Southern Pacific line. Their share of the business was between 20 and 40 percent, though great variances occurred from year to year.

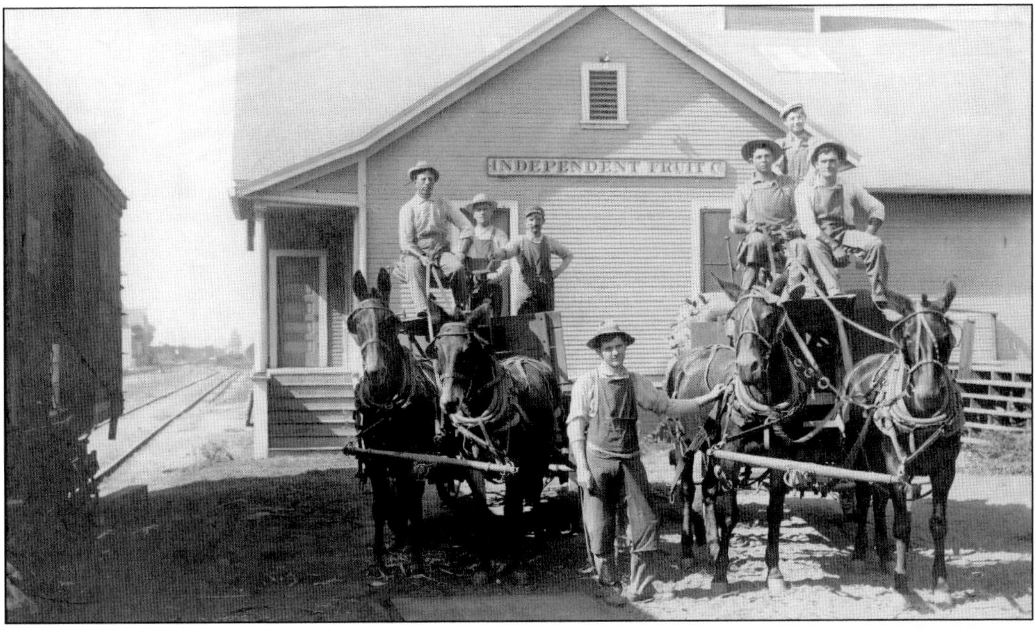

Two mule-drawn wagons are at the side of the Independent Fruit Company's Savoy packinghouse. The workers include T.L. Tinker (left wagon, seated far left holding the reins), Mr. Duke (in the left wagon with Tinker, seated far right), and Matt Matney (right wagon, seated far right). The wagon on the right is filled with oranges in jute sacks, which were used to carry oranges to the packinghouse before field boxes became standard.

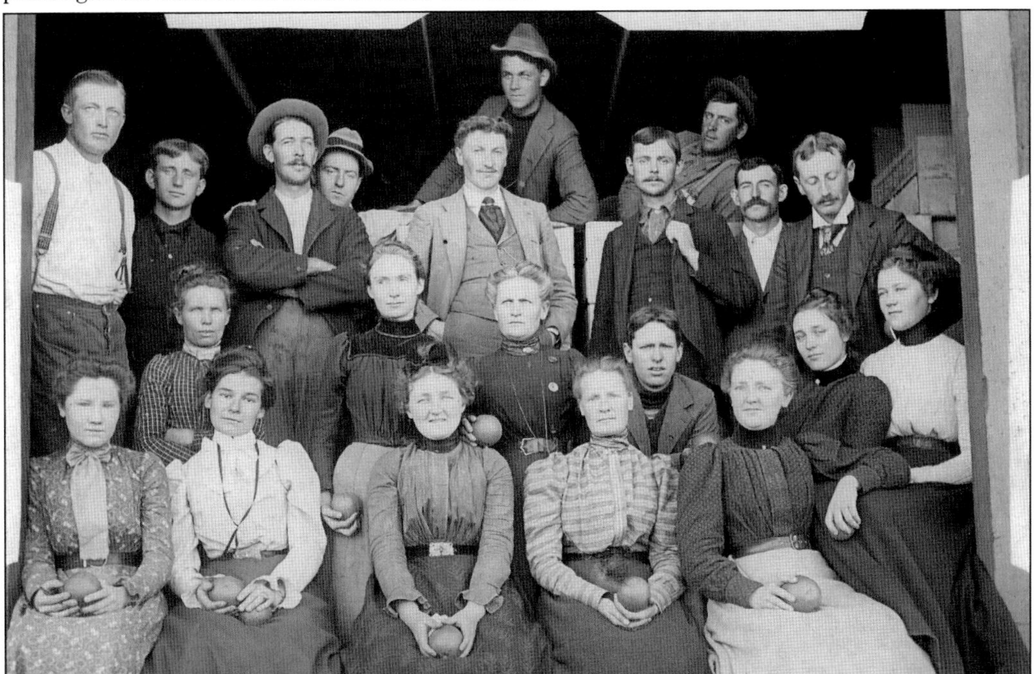

This is a photograph of a small orange packinghouse crew dressed in their best, possibly for a celebration. The ladies are holding oranges. The manager (third row, fifth from left) is wearing a three-piece suit and a silk tie.

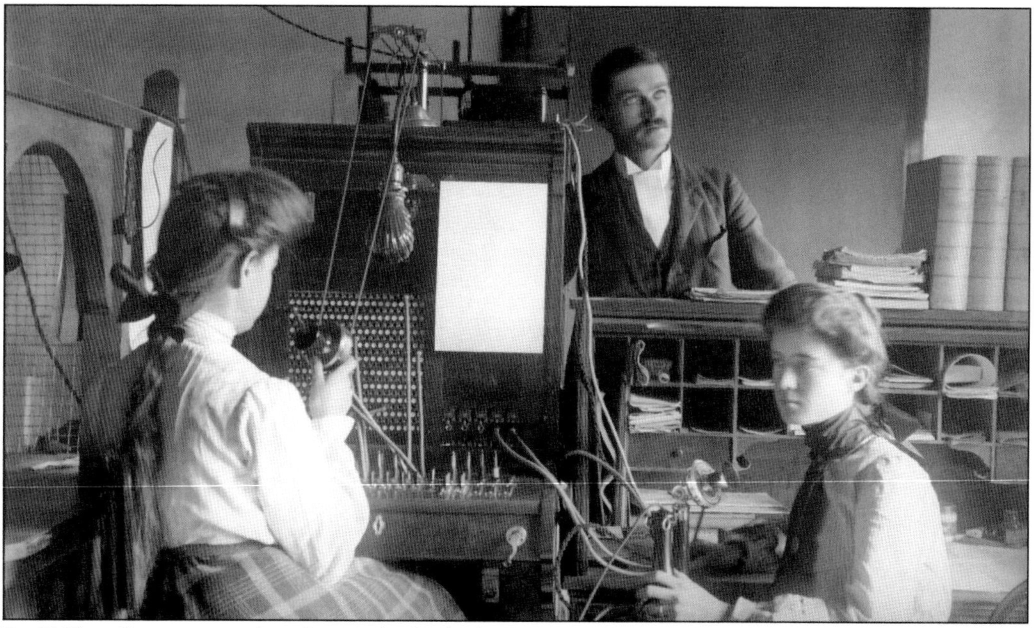

By 1895, valley citrus packinghouses had telephone service from the Los Angeles Sunset Telephone Company. The service was so poor that John O. Houser and P.T. Spencer started the Home Telephone Company, the second local telephone company in California. On September 4, 1902, it began operating with 72 subscribers in Azusa, Covina, and Glendora. Pictured are, from left to right, Josephine Sumpf, Charlie Crawford, and Lola Coolman.

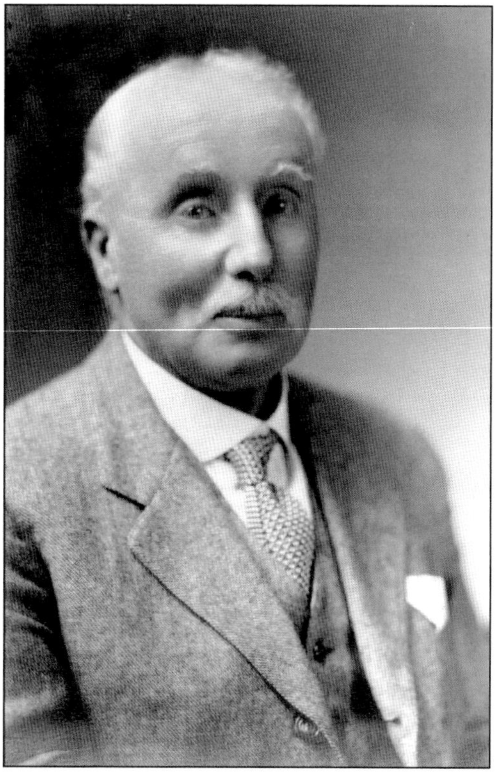

After purchasing 40 acres of the Rancho San Jose, William Bowring planted citrus and built a home surrounded by beautiful gardens. The Bowrings were one of the first families to build a tennis court. Tea and tennis tournaments became very popular in the valley. On March 13, 1902, Bowring invited eight ranchers to form the Charter Oak Citrus Association to join the California Fruit Exchange. (Courtesy of San Dimas Historical Society.)

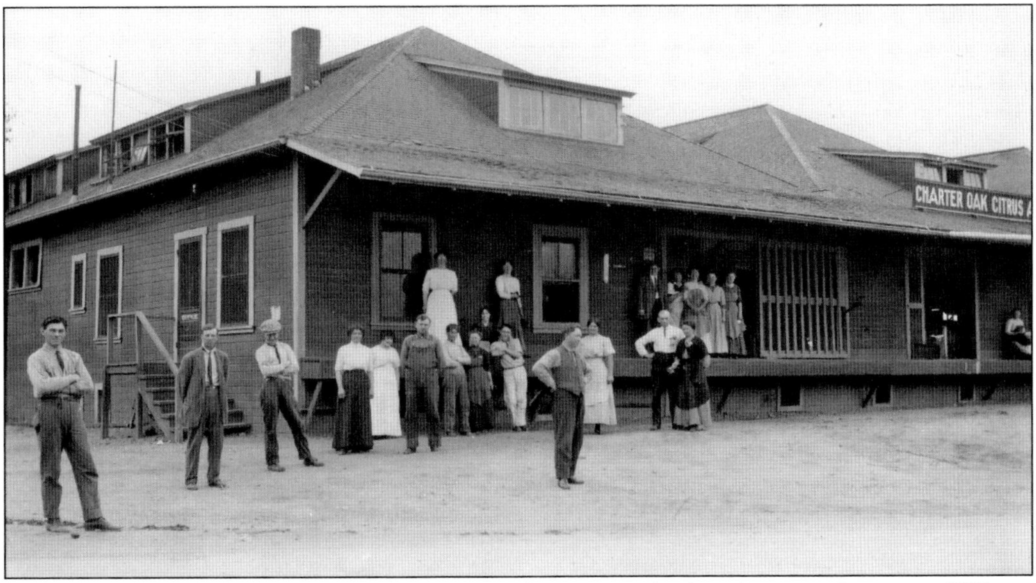

The Charter Oak Citrus Association was formed in 1902. The founding members were William Bowring, H.C. Mace, I.R. Reacon, Arthur Bowring, B.M. Given, W.E. Kent, S.A. Stowell, William Crook, and P.C. Daniels. Here, workers are seen in front of the tidy Charter Oak packinghouse exterior.

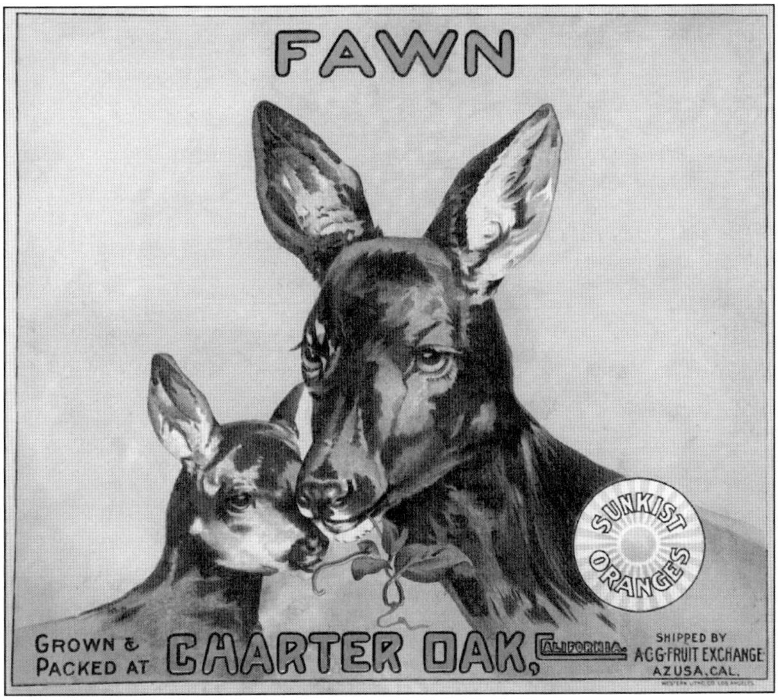

The Fawn label was printed for the Charter Oak Citrus Association by Los Angeles Western Lithograph Company. This one includes the Sunkist medallion. ACG Fruit Exchange Azusa was the shipping agent. Charter Oak was a large citrus-growing area east of Covina, but it never incorporated as a city.

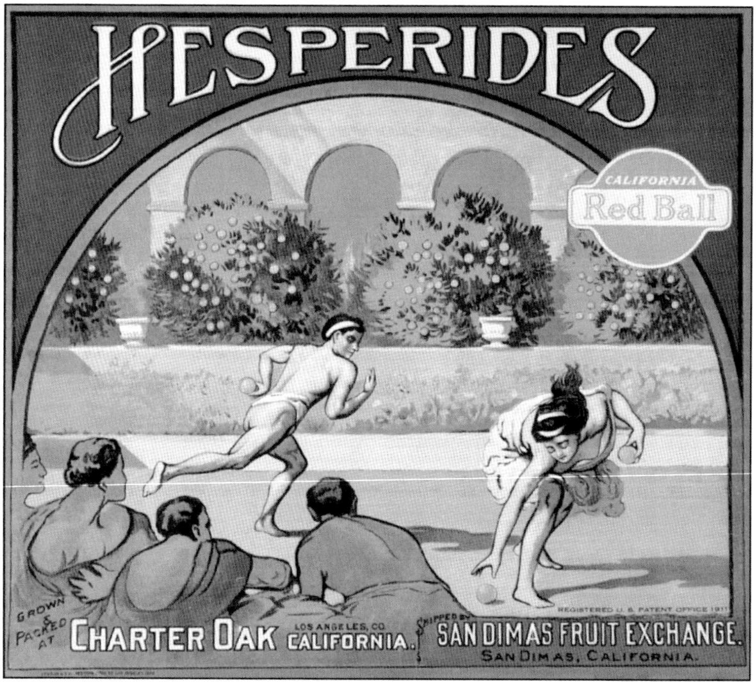

According to Greek mythology, Hesperides, the land of the golden apples, was an island in the far west at the borders of the Western Ocean. The golden apples, wedding gifts for Zeus and Hera, were in a garden guarded by Ladon, the sleepless dragon. The 11th labor of Hercules was to steal the golden apples from the garden. This label is a reference to the story. It was printed by Los Angeles Western Lithograph Company for the Charter Oak Citrus Association.

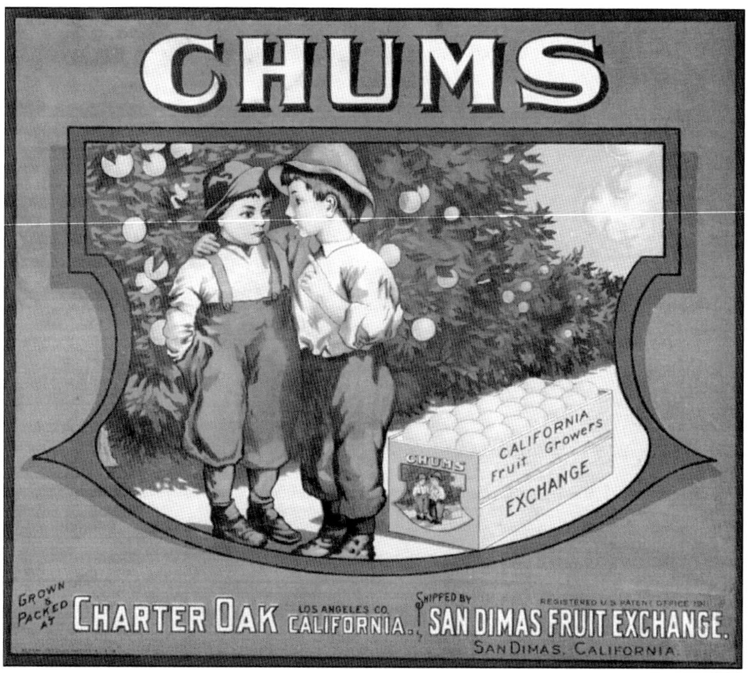

The Chums crate label was printed for Charter Oak Citrus Association by Los Angeles Western Lithograph Company. California Fruit Growers Exchange appears on the side of the packing crate. The San Dimas Fruit Exchange shipped Hesperides- and Chums-labeled oranges.

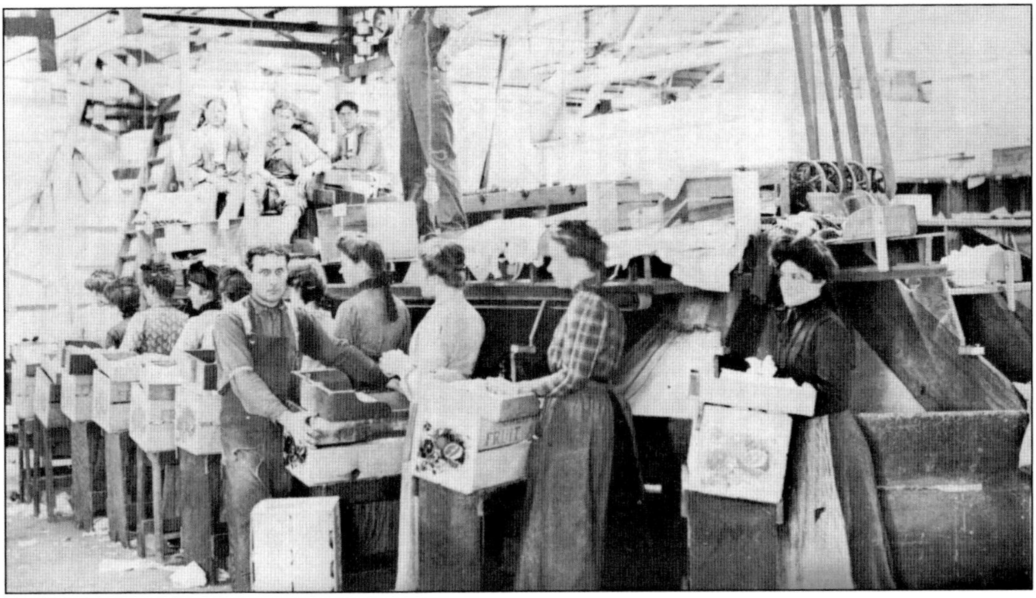

Here, employees of the Stanton and Hendrix independent packinghouse in Covina are shown in the midst of their workday. The female workers are packing Foothill Pansy Brand oranges.

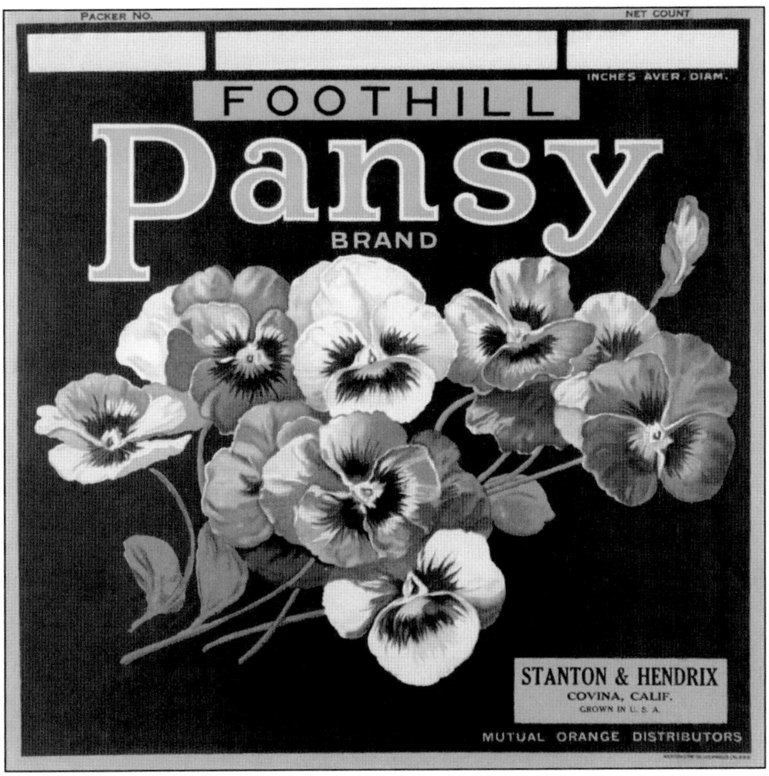

The Foothill Pansy Brand label was printed by Los Angeles Western Lithograph Company for Stanton and Hendrix. Mutual Orange Distributors was the shipping agent. The flower designs on citrus labels were also used in seed catalogues printed by the same companies.

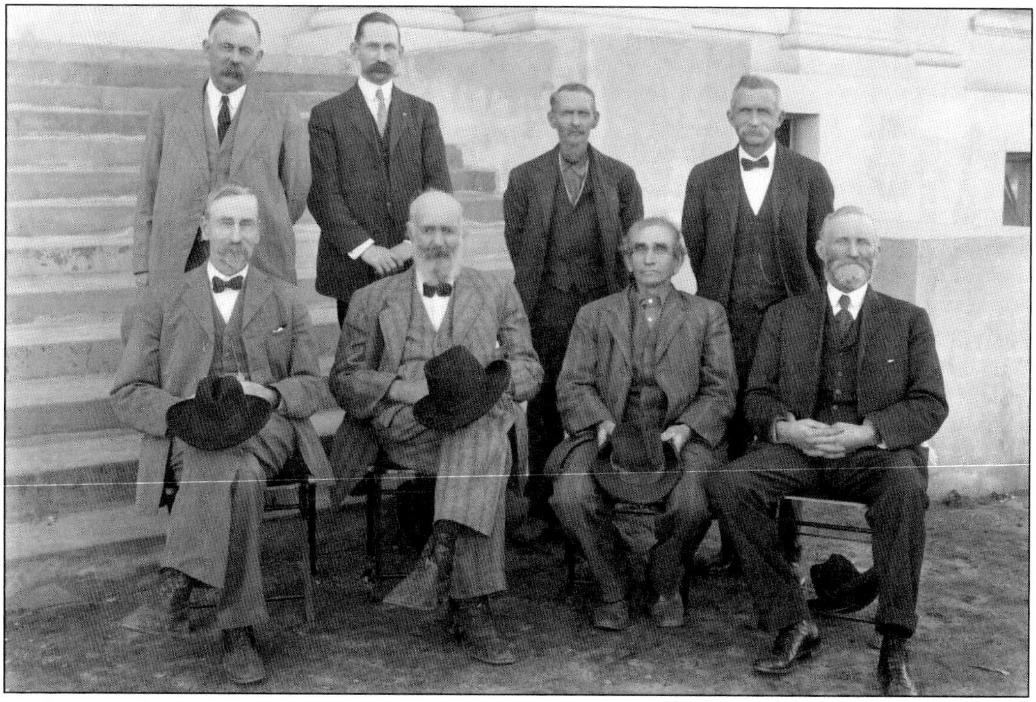

In this rare photograph from 1909, citrus ranchers who had farmed in the Covina Valley for 30 years gather for the dedication of the new Covina High School building. Seated in front of the entrance are, from left to right, (first row) Thomas Francis Griswold, Claiborne Vaughn, Charles Coffman, and Daniel Reichard; (second row) James Elliott, William Merton Griswold, Herman Baldridge, and Ralph Pollard. (Courtesy of Elliott-Milliken collection.)

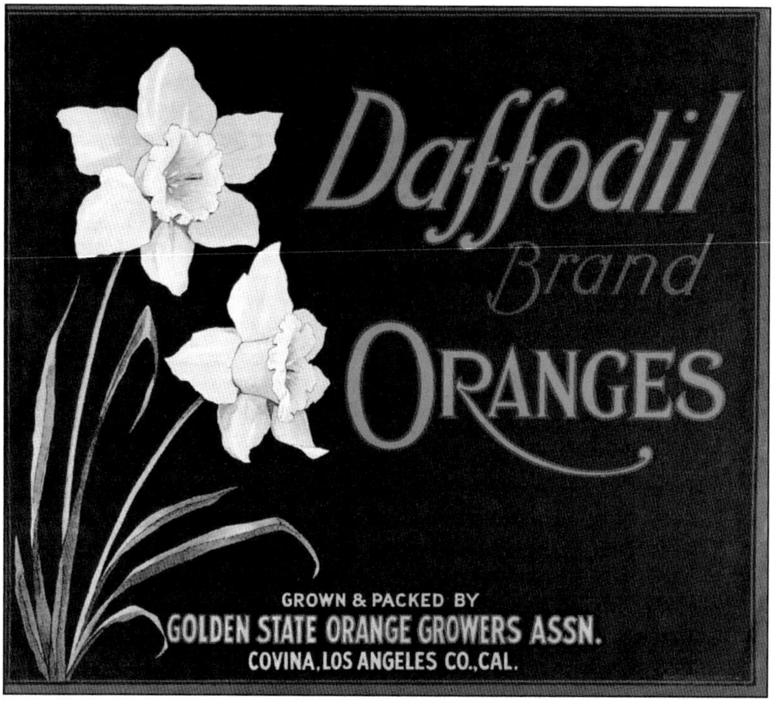

The label for Daffodil Brand oranges was printed by Los Angeles Schmidt Lithograph Company for the Covina Orange Growers Association, which belonged to the California Fruit Growers Association, and for the Golden State Growers Association, which was a Covina independent packinghouse. The Daffodil Brand was listed as a temporary label for Golden State.

The Gold Medal Washington Navels label was printed for another independent Covina packinghouse, Golden Orange Groves, Inc. The label reads, "California's Best Pure Gold Oranges."

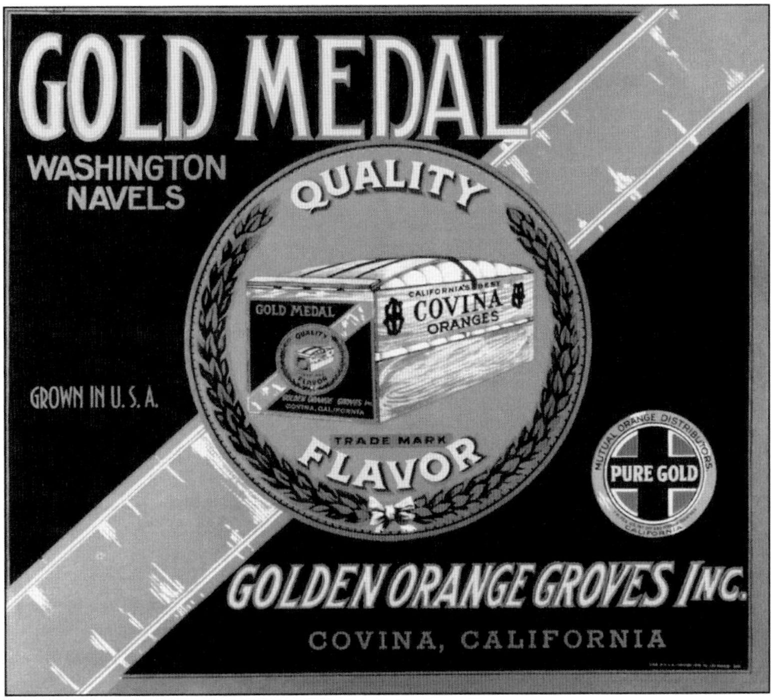

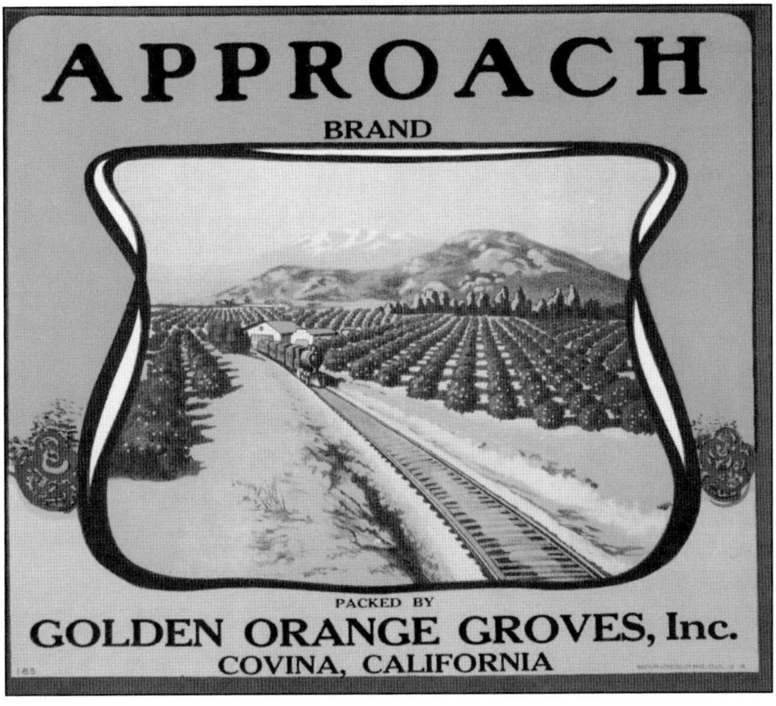

The label for Approach Brand oranges was printed by Los Angeles Western Lithograph Company for independent Covina packer Golden Orange Groves, Inc. The view, however, is actually of the Glendora foothills and the stylized freight train would be a Santa Fe train on its way to San Bernardino.

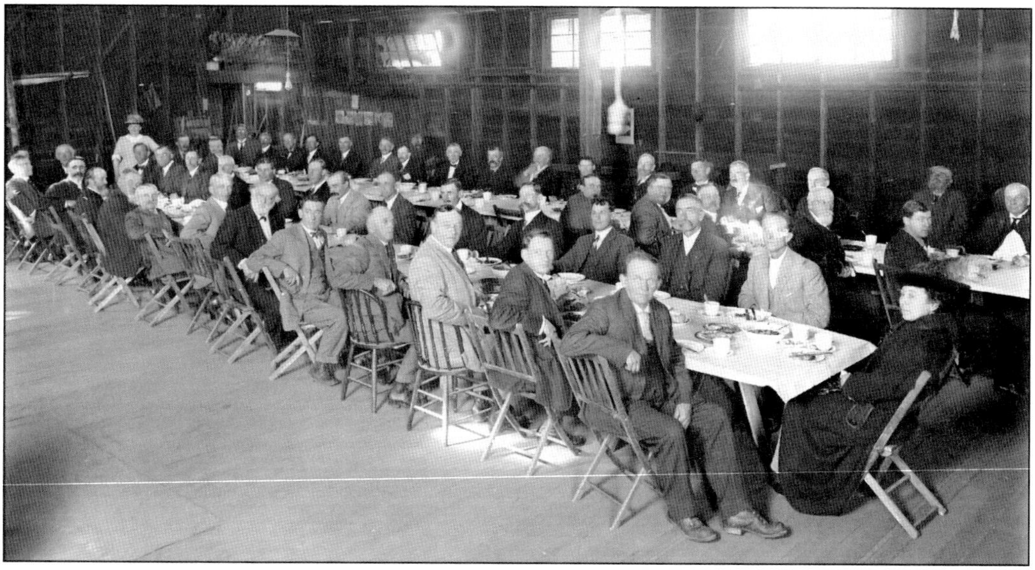

The Covina Citrus Association hosted a stockholder banquet in 1912. After the Southern California Fruit Growers Exchange was incorporated, ranchers contracted with their local packinghouse association, which belonged to a district exchange. The exchanges were represented by directors who each had one vote. William Merton Griswold was the director representing the Covina Fruit Exchange. Anton Kerckhoff was president of the local exchange, and E. Eyman Huff was the manager.

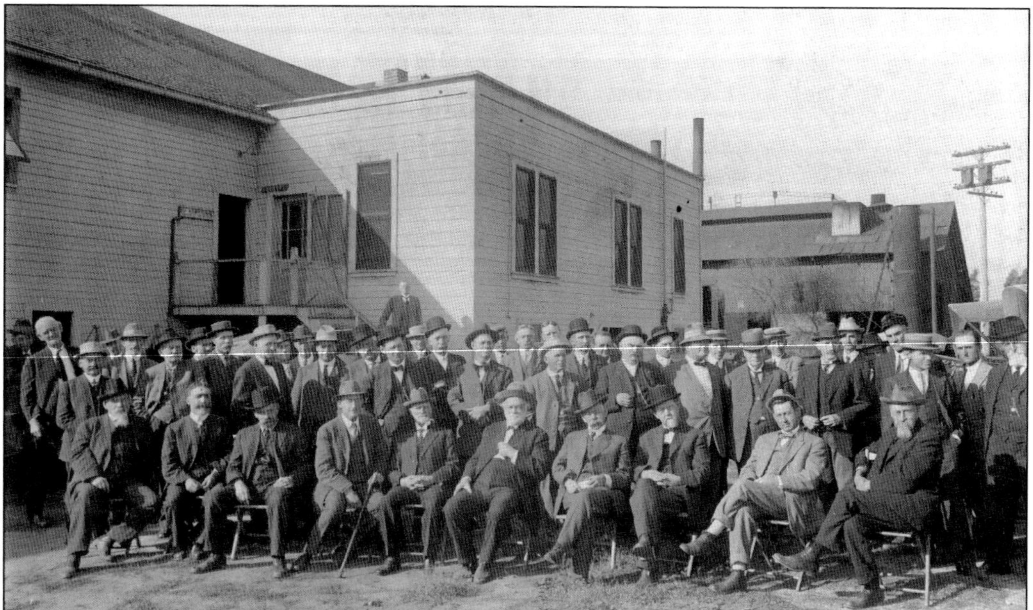

Here, Covina Citrus Association stockholders gather for their annual meeting in 1912. Manager George Anderson is standing on the stairs. Seated are, from left to right, Jacob Maechtlin, J.L. Motter, Jacob Brunjes, W.L. Watts, C.S. Beardsley, W.A. Lee, James Elliott, B.F. Edwards, S.R. Headley, and Thomas Griswold. The Covina Citrus Association had women stockholders—in 1918, proxies were sent by 13 female stockholders—but it is unknown why women were absent from the photograph. Both men and women participated in the Farmers Club.

Five

Maintaining Commercial Citrus Groves

The first waves of valley settlers were primarily farmers, and many were Civil War veterans. Glendora pioneer William Bryant Cullen lost his right arm in the war, and Covina pioneer George Reynolds lost the use of an arm saving his commanding general in the battlefield, for which he was awarded the Gold Metal of Honor (later renamed the Congressional Metal of Honor). These farmers left harsh winters in the East and the Midwest to start a new life in California. When water became available, they found the valley's rich alluvial soil would produce abundant crops of fruit, berries, and vegetables. Farmers raised cows and chickens, and with the exception of salt, flour, sugar, coffee, and tea, pioneer farms were self-sustaining.

Later waves of settlers brought many who had no farming experience at all. Only one community founder from the later group of settlers, John Swift Phillips, had farming experience. Jonathan Sayre Slauson was a banker and real estate investor, and Chicago native Charles Dexter Whitcomb was an inventor and an industrialist. For the most part, they all started on the same page when it came to growing citrus trees.

Phillips strongly believed in the future of citrus. He started selling land to buyers who were planning to come to California and wanted a citrus grove waiting for them. He arranged for valley nurserymen, like W.R. Barbour, to develop and tend to citrus groves for absentee owners. Gradually, as the ranchers who were growing other crops realized that citrus made more money, they tore out deciduous fruit trees and berry vines and planted citrus. The new citrus ranchers eagerly sought information on how to develop their crop. They joined the Covina Valley Farmers Club and the Southern California Pomological Society. The University of California agricultural extension program was just beginning, and in March 1909, a special demonstration train created by the university in cooperation with the Southern Pacific Railroad arrived in Covina. On board the train to give lectures and demonstrations were Dr. B.I. Wheeler, president of the University of California; E.J. Wickson, director of the university experimental station; and T.C. Clarke, superintendent of the university extension department of agriculture. Ranchers and their families came from all over the valley to see the exhibits.

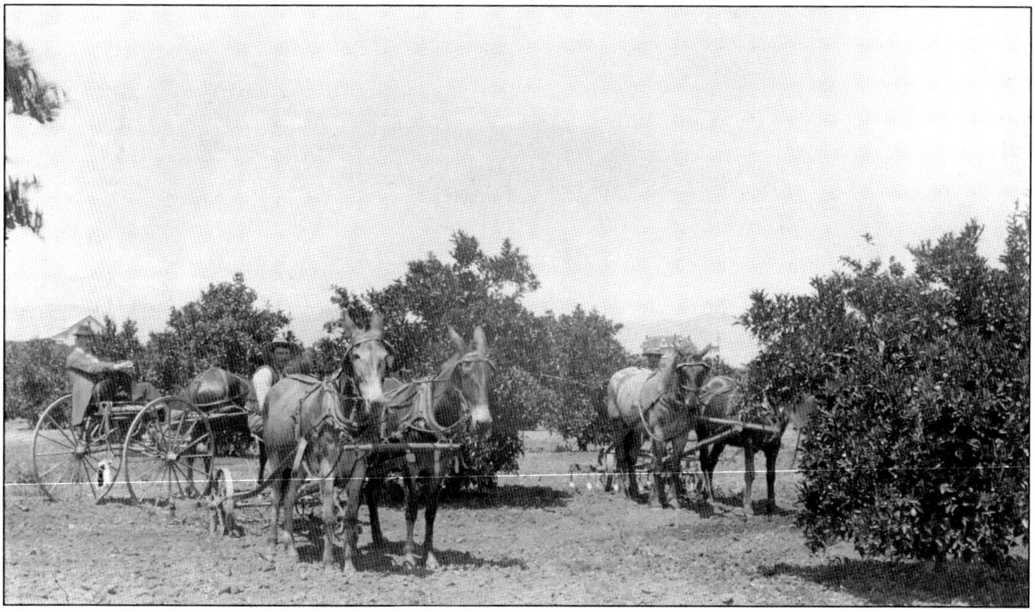

Pictured from left to right are Frank Cartzdafer, Sam Glaze, and Sherm Glaze. Cultivating and furrowing are being done here in the Ruddock grove to prepare it for irrigation. Ranchers kept teams of mules or horses to pull the cultivator. Local blacksmiths and wagon makers developed variations on the spring-tooth harrow, particularly for use in citrus groves. Robert Baldridge designed one with two adjustable scraper wings suspended on either side.

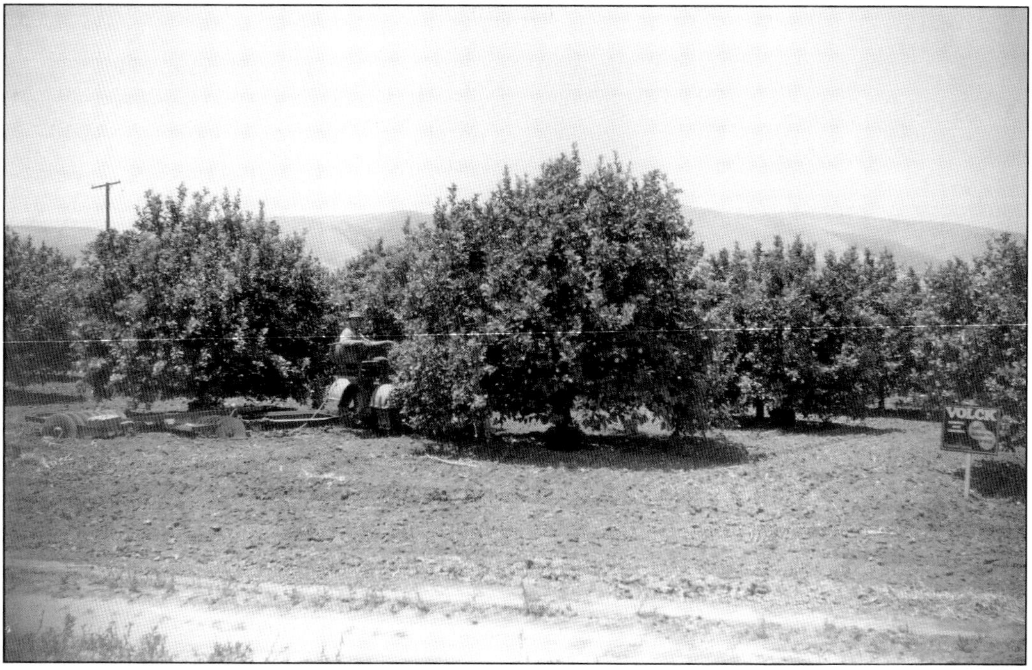

Small tractors, such as the one pictured, replaced mules and horses. Kenneth Avery, a local rancher, invented one of the first tractors, the baby Creeping Grip tractor, for walnut and citrus groves. After World War I, Fordson tractors delivered from Detroit for $497.71 replaced the local models. Buying a tractor was much cheaper than feeding horses and mules.

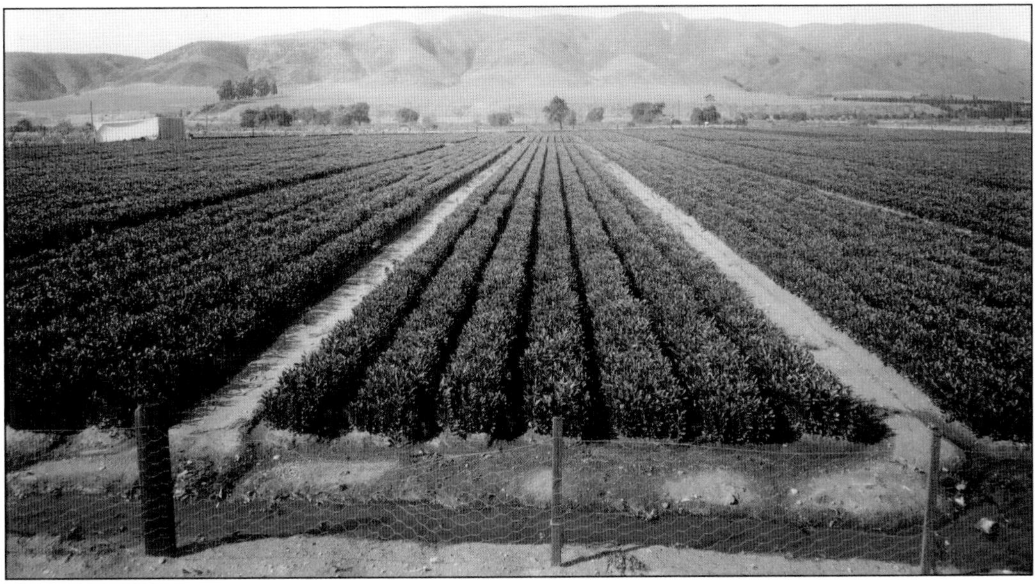

Valley citrus trees were propagated in field nurseries like this one south of Glendora. Seeds planted in the spring were usually ready to bud by fall. They were planted two to three inches apart in one-inch rows. The T bud method was used to bud citrus. Fall buds were unwrapped six to eight weeks after budding. The fruit size of the Washington Navel and the Valencia depended on the root stock.

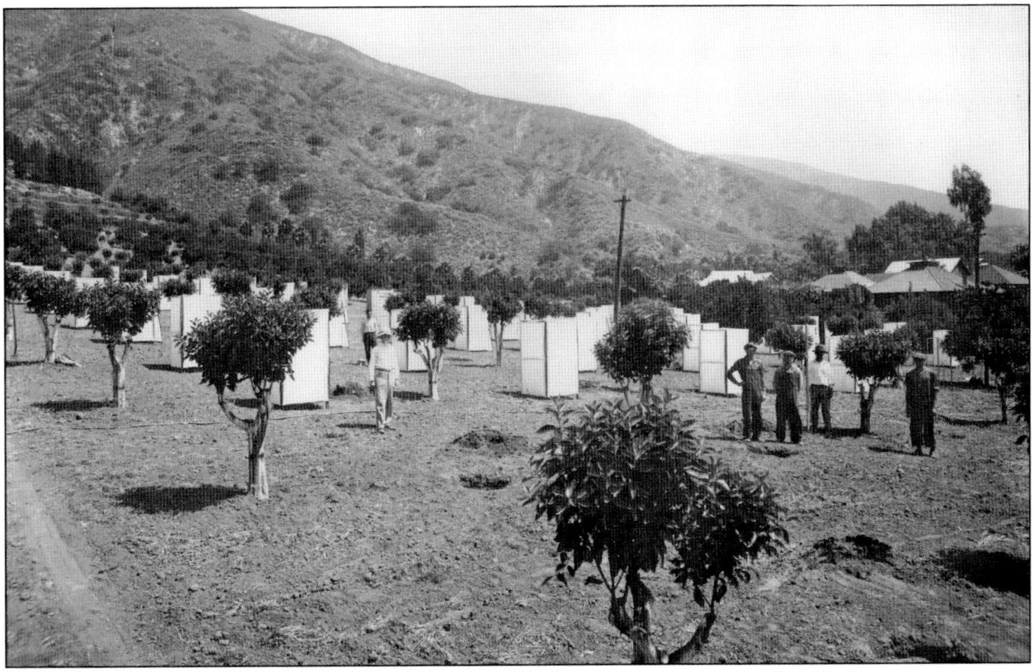

This photograph illustrates the last step in establishing a grove. Young budded citrus trees that were balled and burlaped in the nursery are being planted in a new grove south of Glendora Hills. The man in the center of the picture with the white beard is Verne Jobe. The largest Washington Navels were grown on sour orange rootstock. The dwarfing trifoliate orange rootstock produced the largest Valencias.

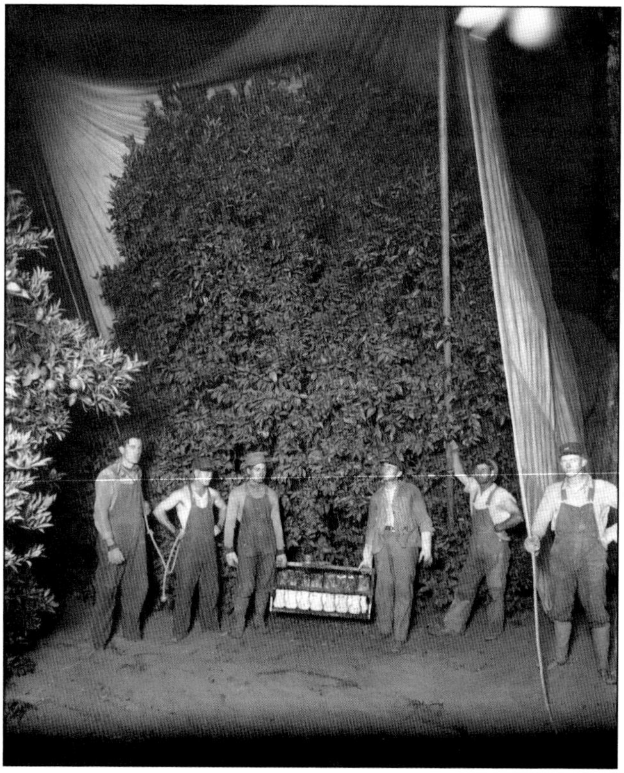

A fumigating crew in the Covina area is preparing to treat a grove. Two men, known as pullers, placed the canvas over the tree. Three other men set two poles in the ground and tramped edges of the canvas flat. The men in the center of the photograph are holding the ceramic containers of cyanide gas used to fumigate the trees.

After 1909, spraying and fumigation increased. It was discovered that a combination treatment was best. Fumigation in August and September was combined with spraying in November. A gas-tight, government-issue twill improved fumigation. Several crewmembers are posed behind the cyanide gas containers with a curve dash Oldsmobile that has been made into a truck.

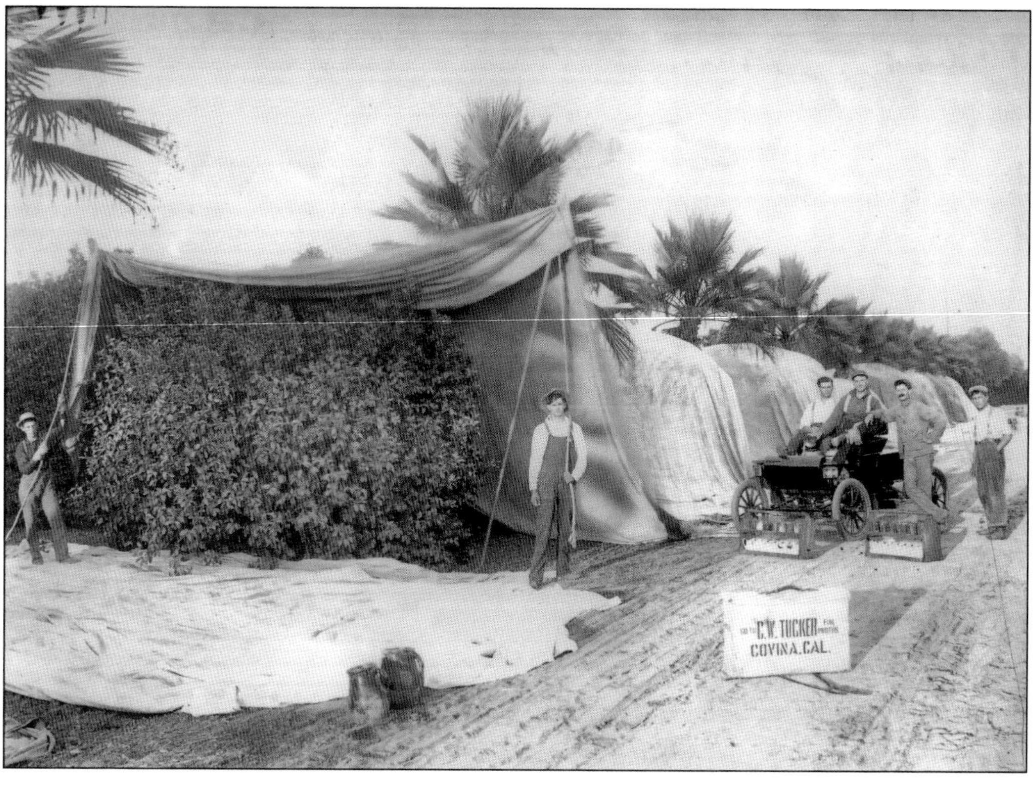

Cullen Motter is operating the cyanide generator that has been pulled to the site by a horse. Fumigating was usually done at night.

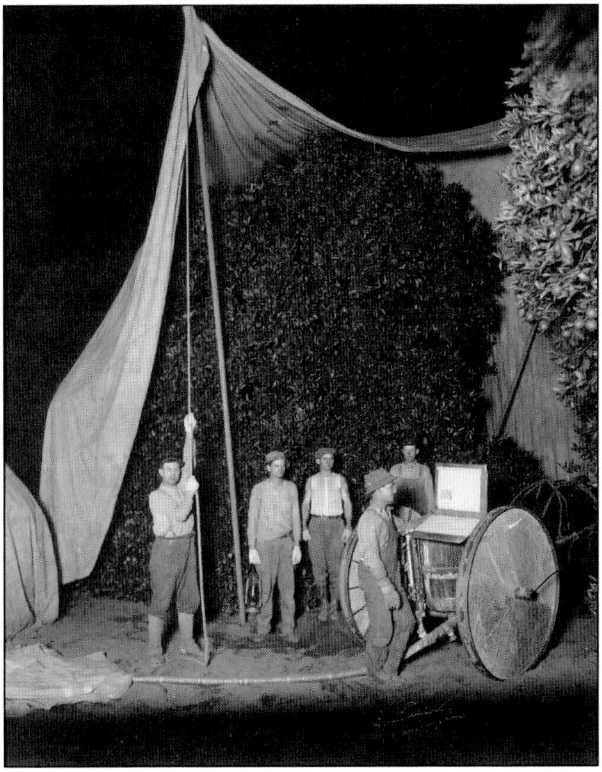

The spraying inspectors are examining the results of spraying a grove. They are wearing jodhpurs and field boots with white shirts, bow ties, and hats. The outfits are representative of what agricultural inspectors and other professionals who worked in citrus groves wore at the time.

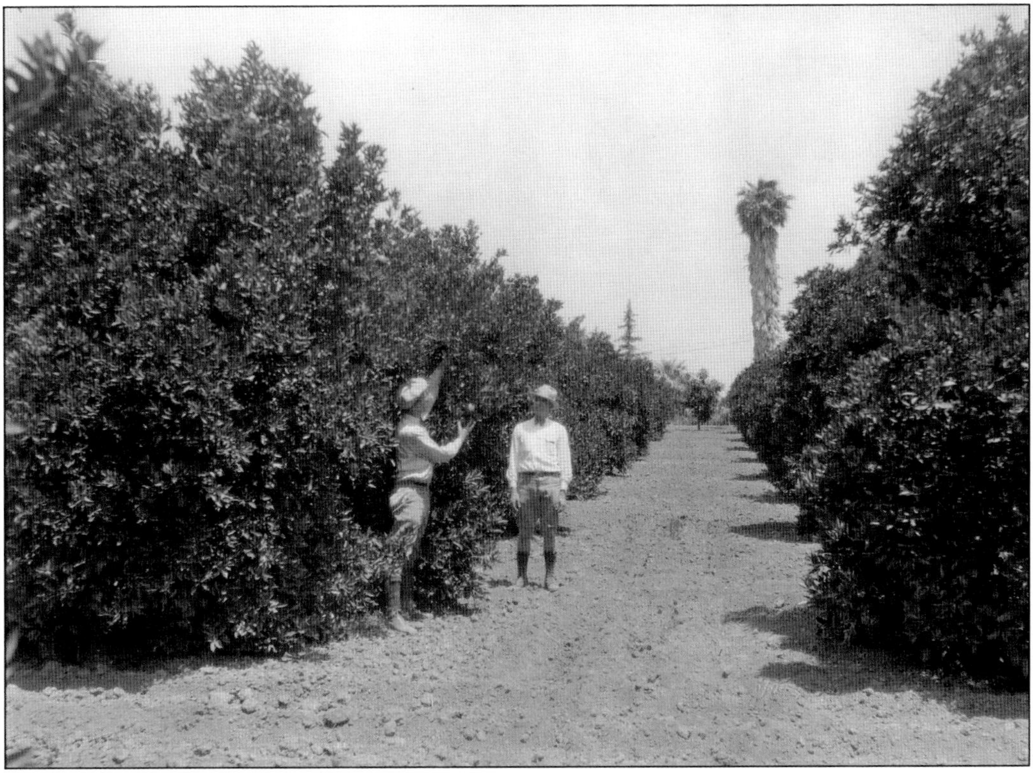

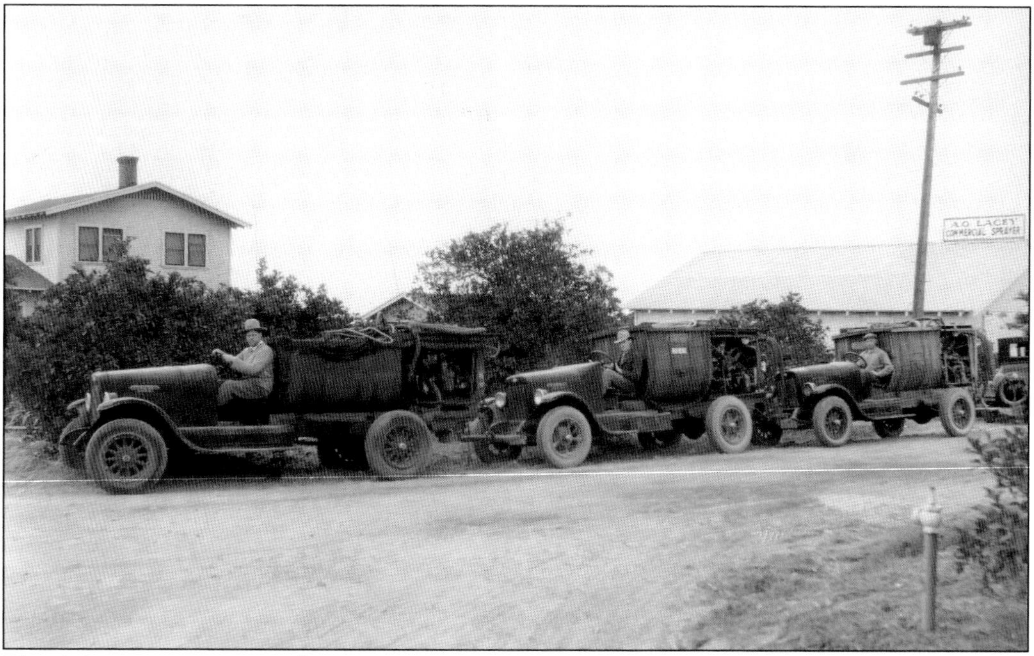

The A.O. Lacey Commercial Spraying Company's trucks are lined up for this photograph. Several Glendora businesses serviced the citrus industry. John Hastie Manufacturing and Engineering Company (formerly Warren-Hastie) provided Pomona Pumps, packinghouse equipment, and equipment for irrigation, pre-cooling, and air-conditioning. Another Glendora business, the Foothill Supply Company, was formed by officers and members of the Glendora Cooperative Fumigating Company in 1923.

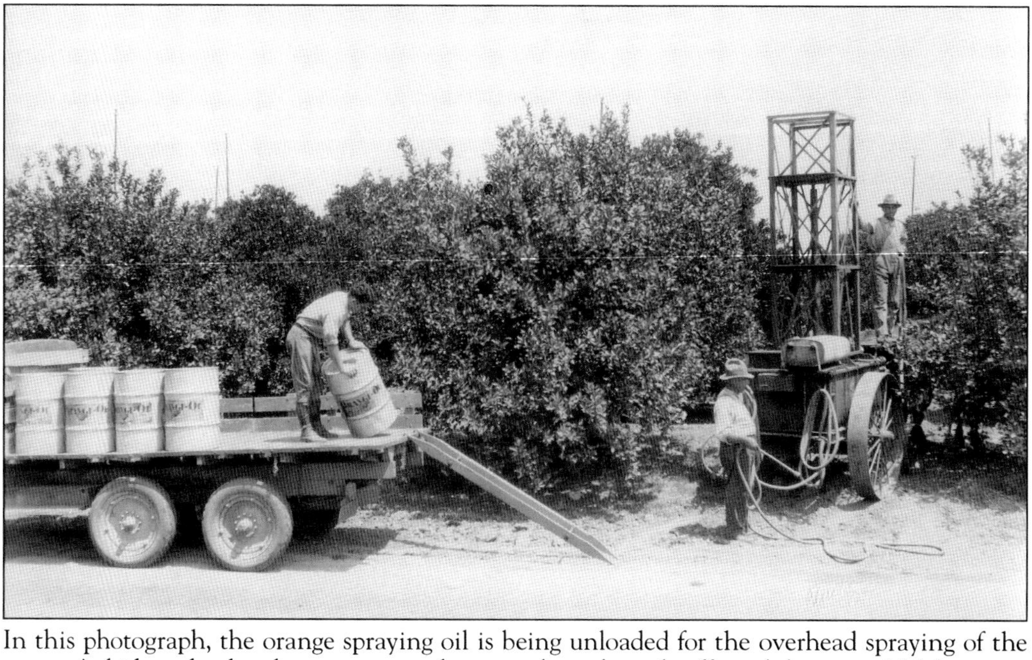

In this photograph, the orange spraying oil is being unloaded for the overhead spraying of the grove. Aphids and red scale were among the pests that adversely affected the crops. Volck orange oil was used. It was a dormant oil that replaced cyanide spraying and fumigating.

These men, at right, are spraying from a rig. In 1888, fumigating and spraying experiments began because Arizona passed laws about California citrus. By 1909, other states began passing laws against untreated California fruit that was arriving diseased or covered with dead scale. These photographs were taken in 1910. Using a small rig that could be driven between the trees, the men below are spraying from the ground.

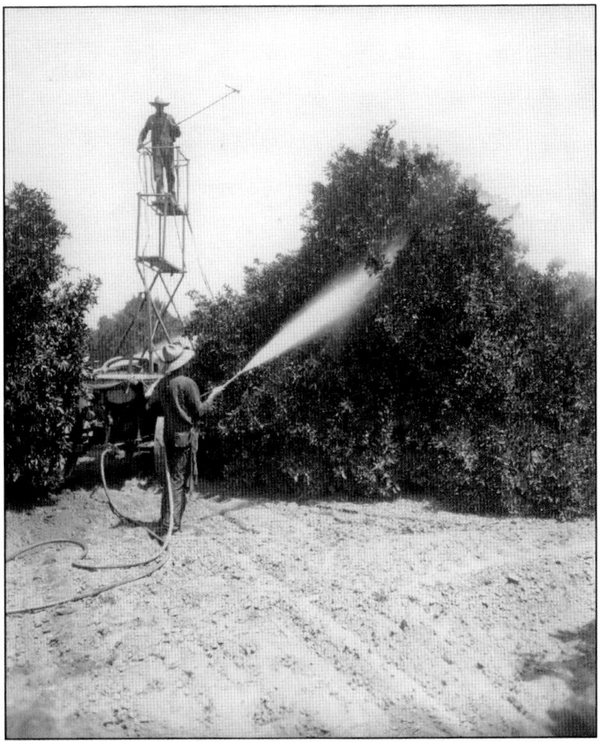

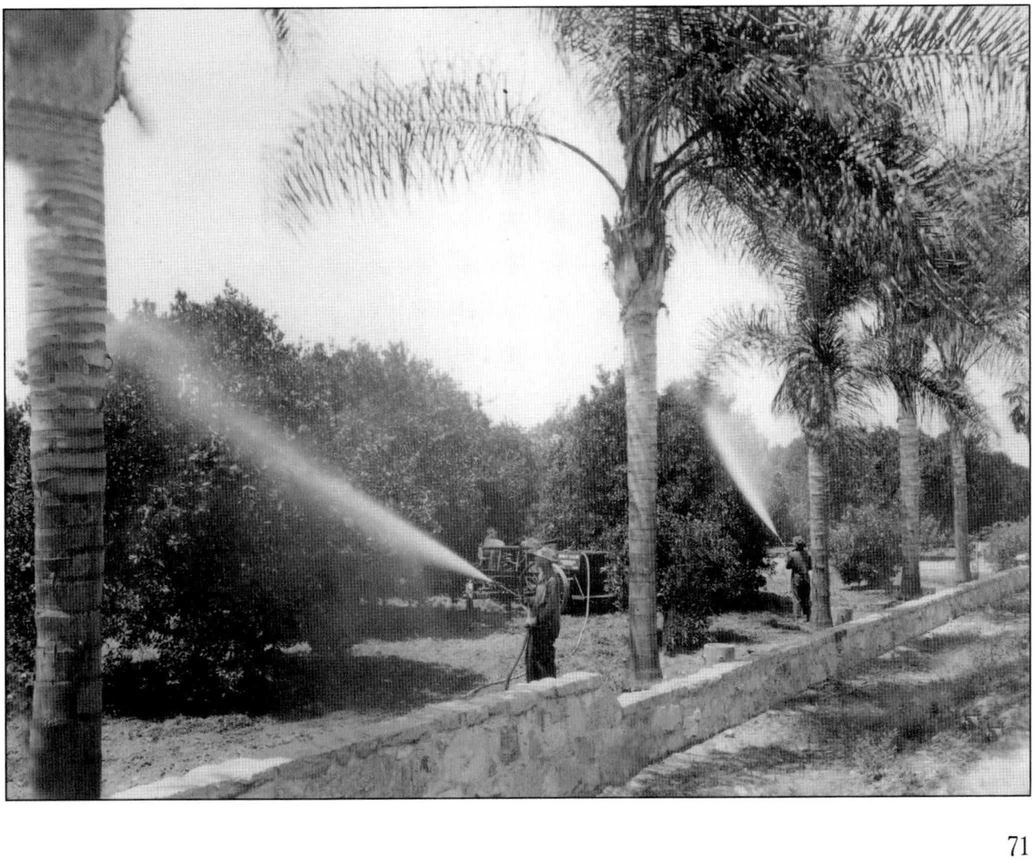

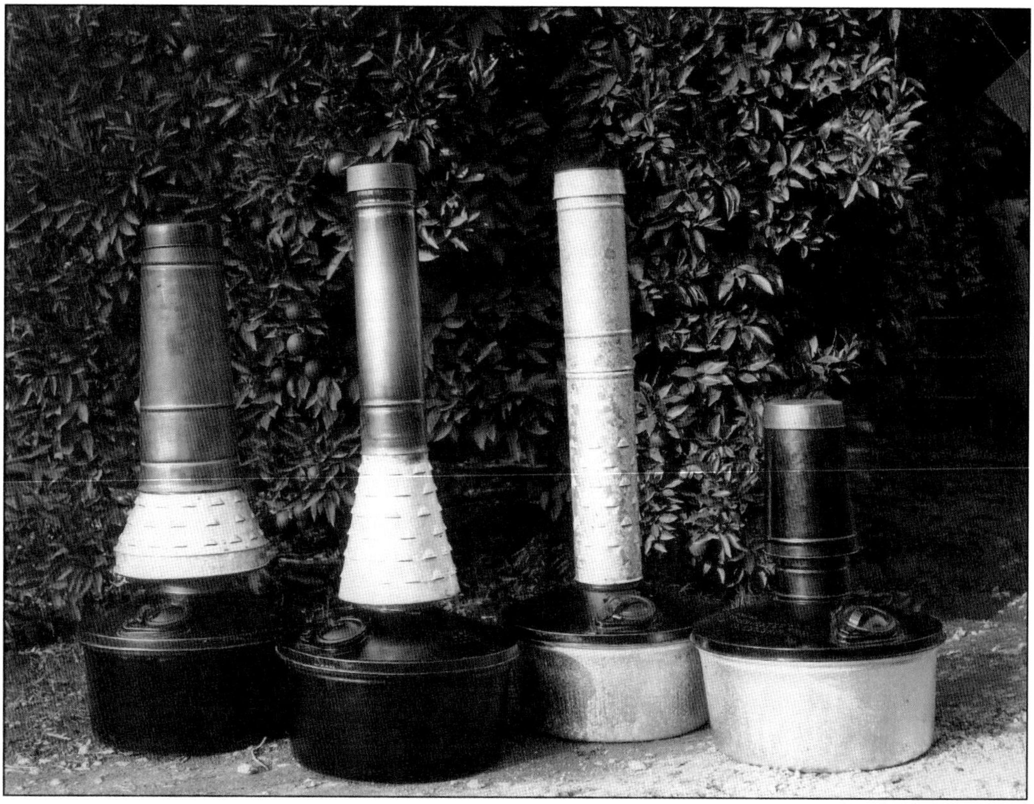

These four orchard heaters, known as smudge pots, were used to keep citrus trees warm when the temperatures dropped below 20 degrees Fahrenheit. The heaters shown include, from left to right, a jumbo cove, a baby cove, a six-inch-tall stack, and a short double stack.

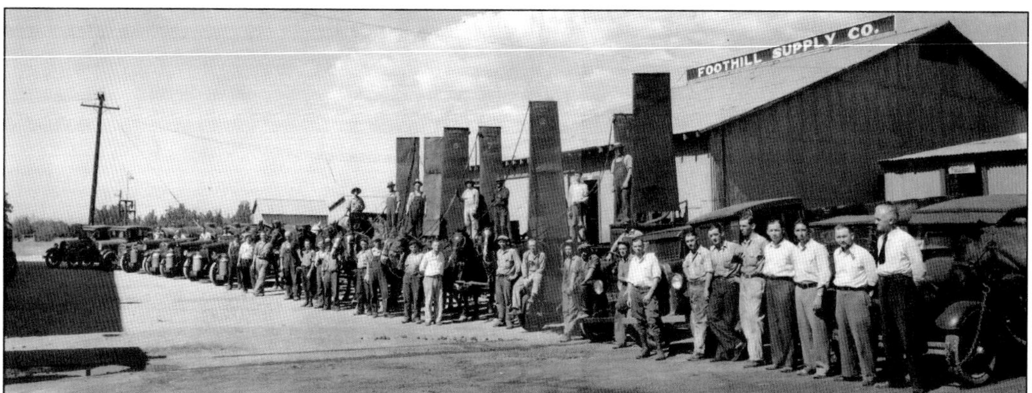

The crews, supply trucks, and spraying tower of the Glendora Cooperative Fumigating Company (Foothill Supply Company) are lined up with the teams of horses and mules that pulled them through the groves. Notice the mule on the right tethered to the car.

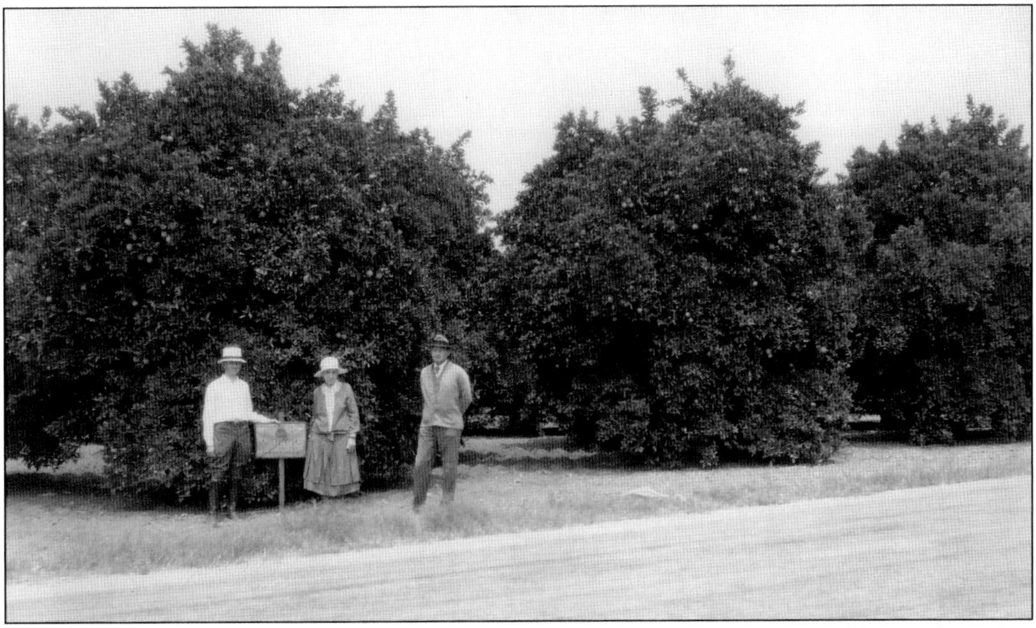

Pictured in front of their grove on Gladstone are George Olmstead (left); Mrs. Olmstead, mother (center); and brother Alfred Olmstead. A Volck sign is posted on the property. Volck, a dormant oil, was sprayed under pressure in late November or early December to control overwintering pests. It was mixed with lime sulfur for control of insects and disease. The Olmsteads were using the latest pest control treatments.

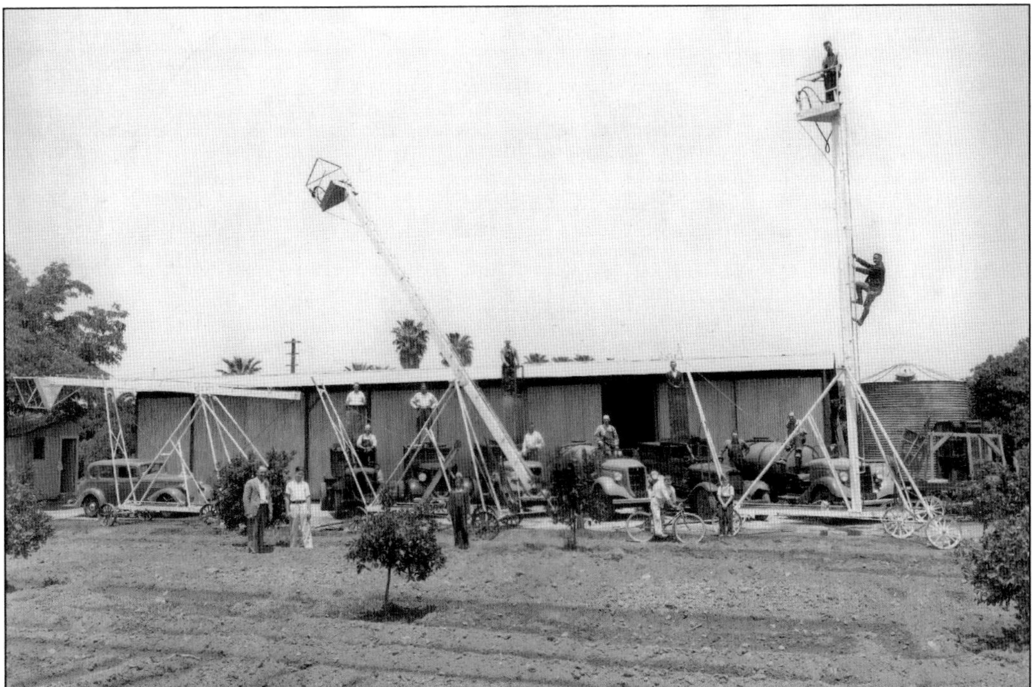

The Covina Citrus Association also had spraying trucks and a different style of tower that could be adjusted in numerous directions.

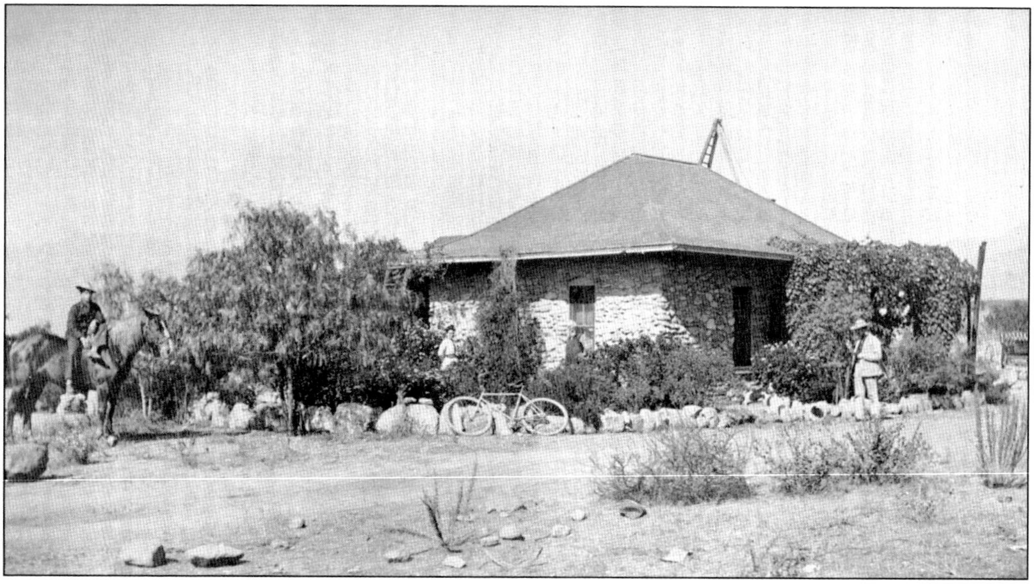

The Columbia Land and Water Company provided well water to Covina's domestic water supply and 1,400 acres of citrus land east of Covina. The company was founded at the turn of the 19th century. John Coolman was president. Pictured is their pumping plant in the San Dimas wash. On the far right is Walter Wheeler of Covina, the accounts keeper for the company. Shortly after this photograph was taken, Walter Wheeler was killed by lighting while hunting in the local mountains.

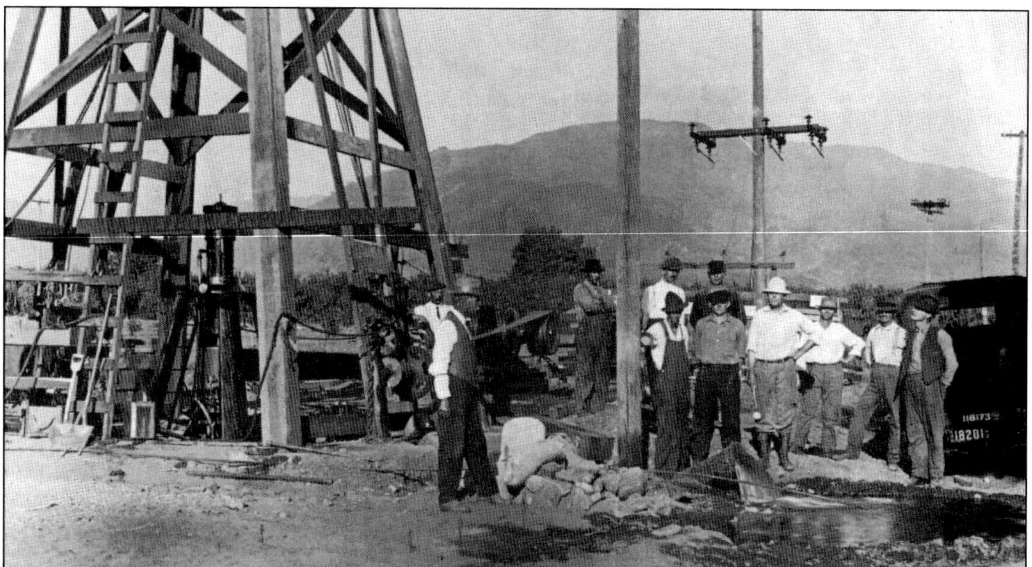

Glendora's first water well, pictured, was completed in 1915. Around this time, all valley communities began to dig water wells for domestic and irrigation purposes. (Courtesy of Glendora Historical Society.)

Six

Golden Years of Valley Citrus

Beginning in the 1920s, the valley citrus industry went through many organizational changes. In 1923, the Glendora Citrus Association and the Glendora Heights Orange and Lemon Growers joined together to establish the Glendora Fruit Exchange. They built a huge packinghouse. According to the *Glendora Gleaner*, "This packinghouse is the largest citrus fruit packinghouse in the world with equipment and facilities good for a pack each season of from 1,500 to 2,000 carloads of 400 packed boxes each." At this time, all the Glendora packinghouses affiliated with the California Fruit Growers Exchange left the ACG Fruit Exchange and shipped through the new Glendora exchange. Also in 1923, the Glendora lemon growers separated from the San Dimas branch packinghouse in Glendora and formed their own association. New independent Glendora packinghouses appeared in 1923 and 1924. Hall-Riser Fruit Company began operating a new plant on North Grand Avenue near the end of Meda Avenue. In 1924, the Mutual Orange Distributors organized the Glendora Cooperative Citrus Association and enlarged the packing plant at Vermont and Ada Avenues.

The two most serious threats to Covina Valley citrus industry were the possible loss of irrigation water and the periodic floods that washed out groves, bridges, utilities, and flooded packinghouses. An organization called the San Gabriel Valley Protective Association, which was backed by Pasadena, Monrovia, Alhambra, Long Beach, Whittier, and 12 other cities, claimed the San Gabriel River water. They involved the State of California in their claim, and the second battle of the San Gabriel River began. Instead of dynamite, the Committee of Nine fought in the courts for over 10 years until a compromise agreement was reached. Morris Dam was one of the results of this agreement. It furnished water for Pasadena and stopped the periodic floods in the valley. During this period, citrus ranchers enjoyed the amenities made possible by their booming economy. Banks as well as savings and loans were available in each community, and they all had opera houses. Valley residents gathered in musical clubs, scientific study circles, and many other civic and social clubs.

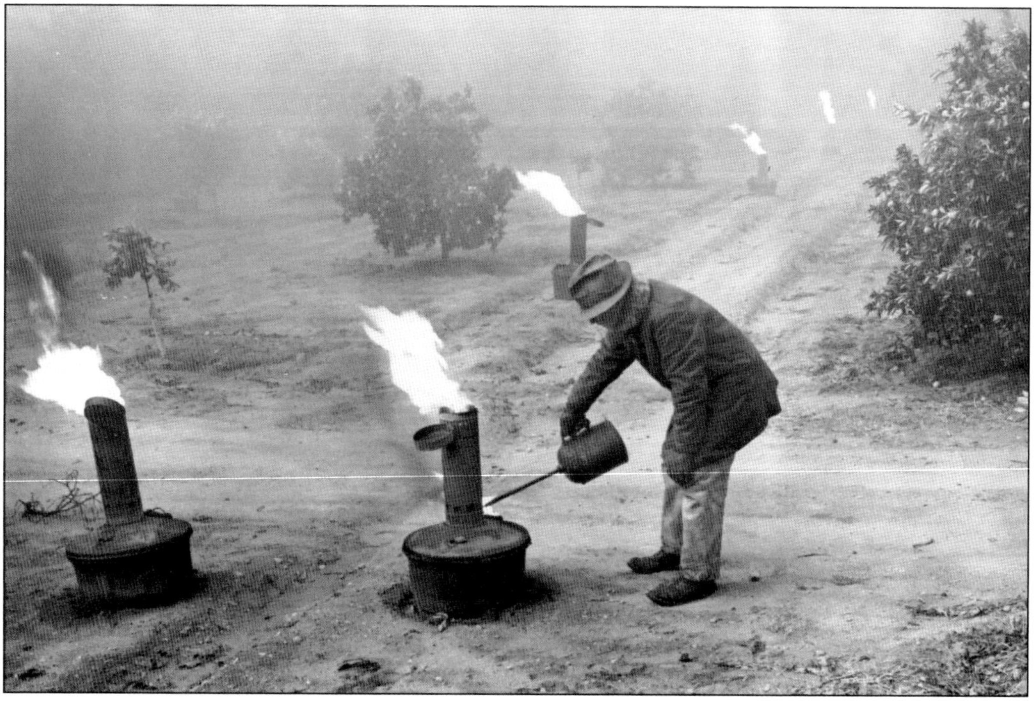

Smudging, a process whereby a smoldering mass is placed on the windward side of the tree to protect it from frost, started in the early part of the 20th century. The earliest ranchers filled paper sacks with shavings and crude oil that burned for 12 hours. Around 1912, various types of orchard heaters appeared. Covina groves were spared frost damage in several major freezes, but in 1913, the crop was severely affected. The Covina Citrus Association purchased smudge pots for ranchers, allowing them to defer payment until after their crops were sold.

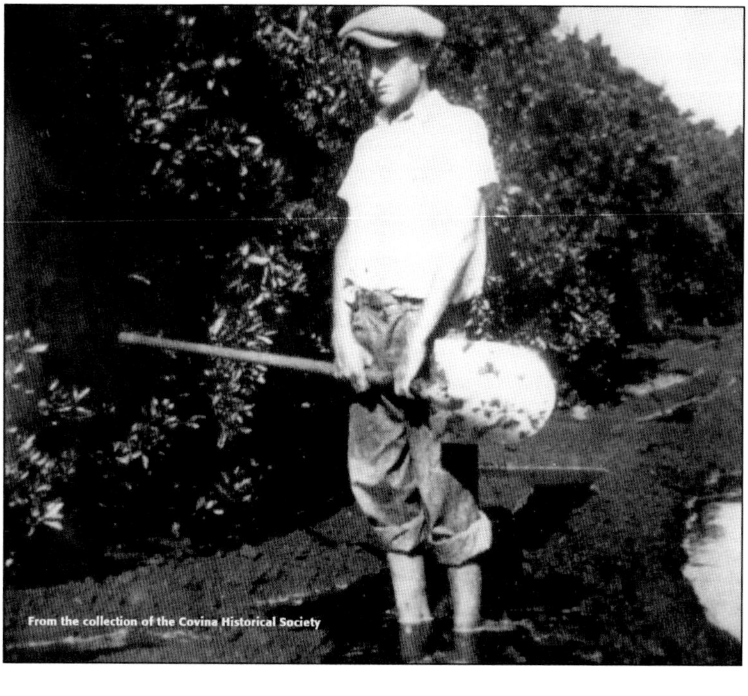
From the collection of the Covina Historical Society

Making furrows for grove irrigation was done by cultivators pulled by horses and, later, by tractors. Cultivators were critical to keeping the furrows open. Young boys like Clair Jobe checked them before irrigating was started. Children walking to school through the groves were careful not to damage them. Furrow systems were used when there was enough of a slope for water to run well. Otherwise, a check system was used.

Here, a Damerel-Allison picker unloads the oranges from his picking sack into a field box. The canvas sacks worked well and did not bruise the fruit.

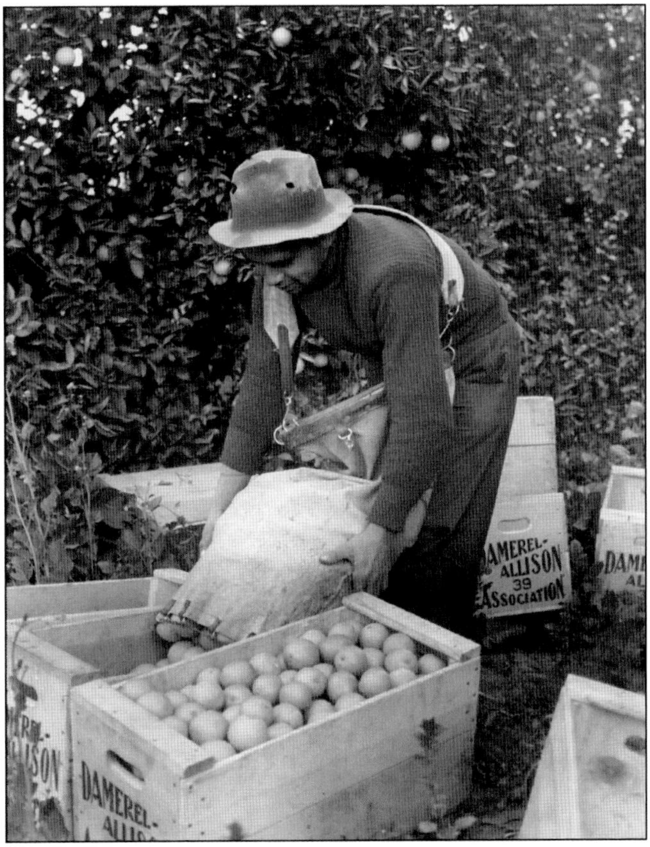

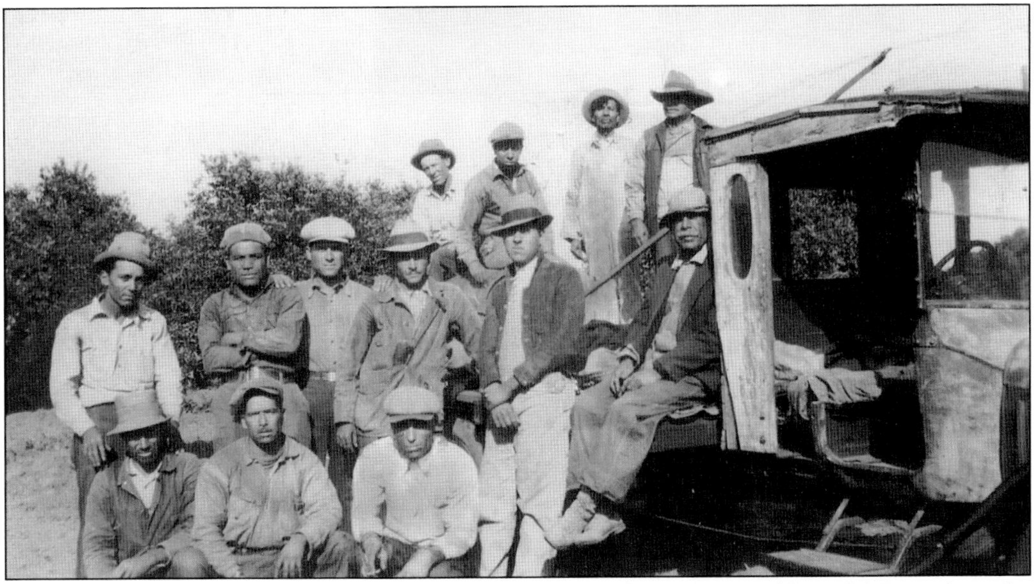

Los piscadores de naranjas (the orange pickers) rode aboard Jim Ayon's old Model T truck to the Hutchinson grove in Covina. Ayon was the *mayordomo* who supervised this Azusa crew. They were picking oranges for the Damerel-Allison packing plant in Covina. An average picker could fill 35 to 50 field boxes a day; a fast one could fill 80. Pickers were paid 40¢ per hour.

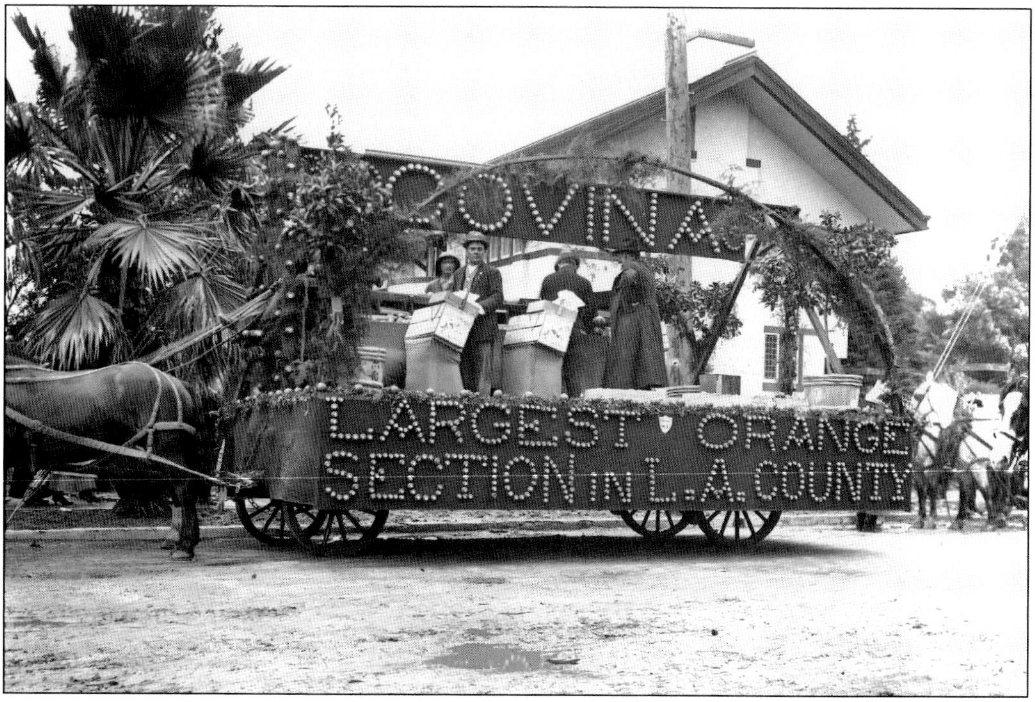

On January 1, 1911, Vernie Houser, Maye Allison, Ruby Keefer, and Marjorie Capon threw oranges to spectators from Covina's second Rose Parade entry. The 1910 float was almost destroyed in a rainstorm, but Clarence Allison and H.M. Houser managed to get it to Pasadena. Covina's first Tournament of Roses appearance was in 1908 when the Covina Independent Order of Odd Fellows marched down Colorado Boulevard.

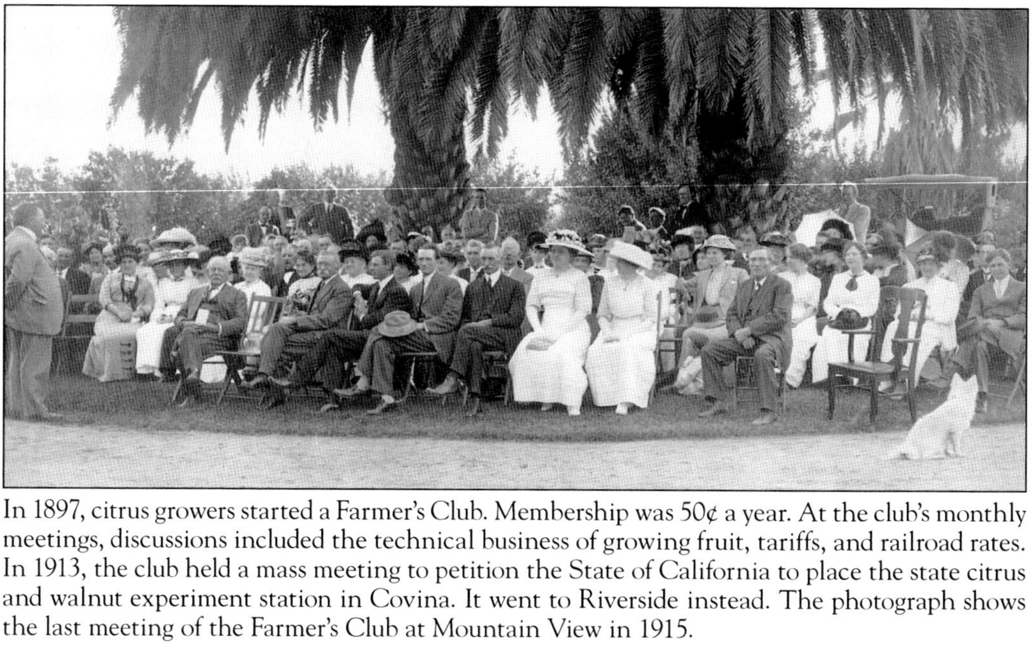

In 1897, citrus growers started a Farmer's Club. Membership was 50¢ a year. At the club's monthly meetings, discussions included the technical business of growing fruit, tariffs, and railroad rates. In 1913, the club held a mass meeting to petition the State of California to place the state citrus and walnut experiment station in Covina. It went to Riverside instead. The photograph shows the last meeting of the Farmer's Club at Mountain View in 1915.

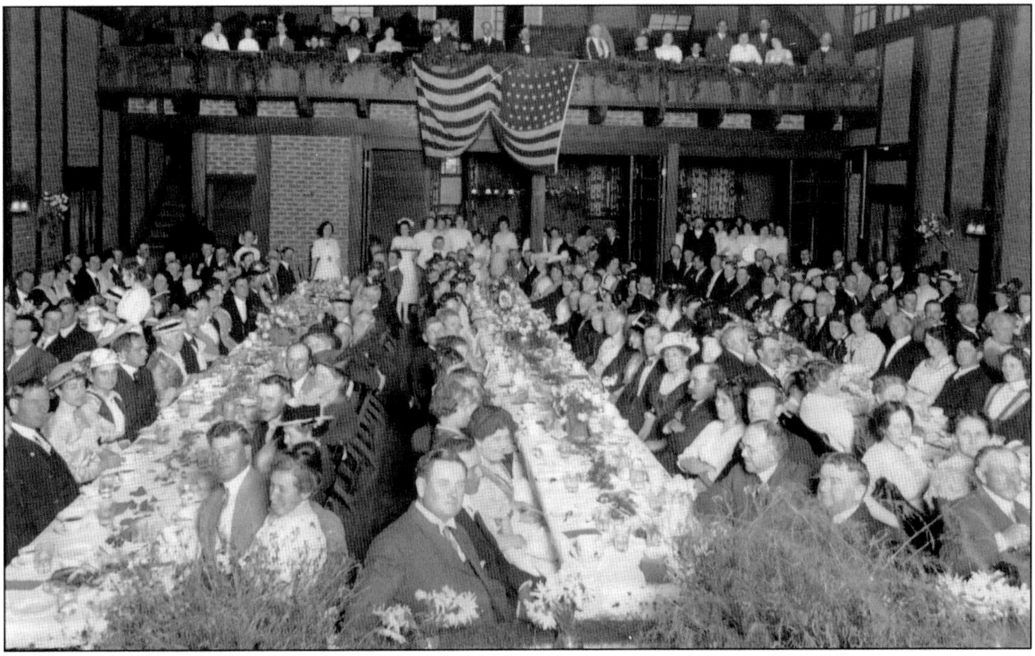

The Covina Orange Club held a banquet meeting at the Covina Woman's Club to organize an advertising campaign promoting the sale of Covina citrus fruit; approximately 250 supporters attended. According to H.A. Miller, president of the boosters, the State of Florida was marketing their inferior fruit to East Coast cities, which was affecting the Covina farmer's sales. The speaker was G. Harvard Powell, manager of the California Growers Exchange.

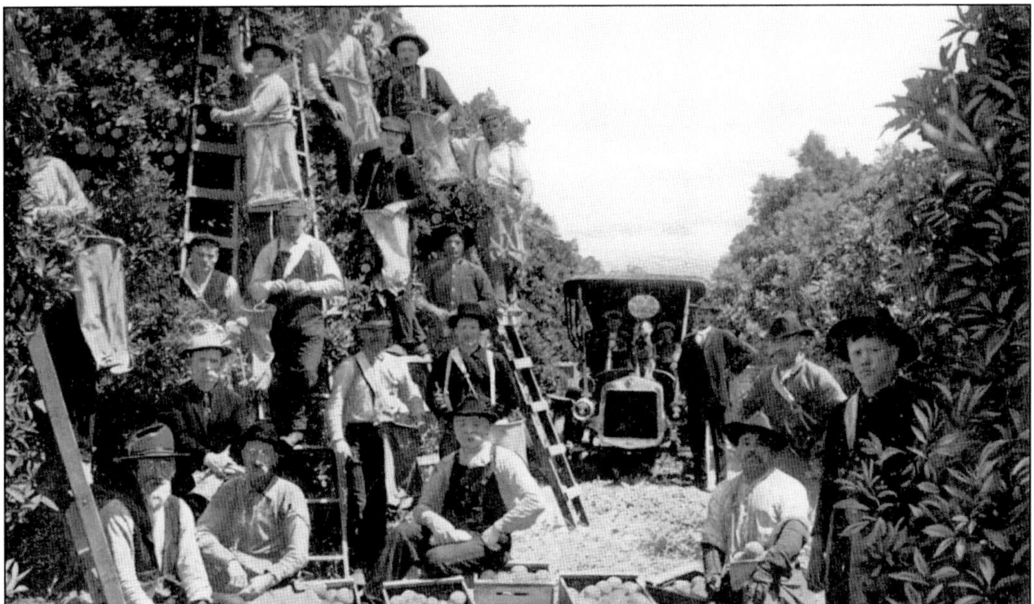

A rancher and his son are posed in their automobile with a Dalmatian perched on the hood. They are in a mature grove with the picking crew and the foreman. By 1895, Covina's population had doubled. Land with lemon, orange, or deciduous trees sold for $350 to $600 per acre, depending on the location and the age of the trees.

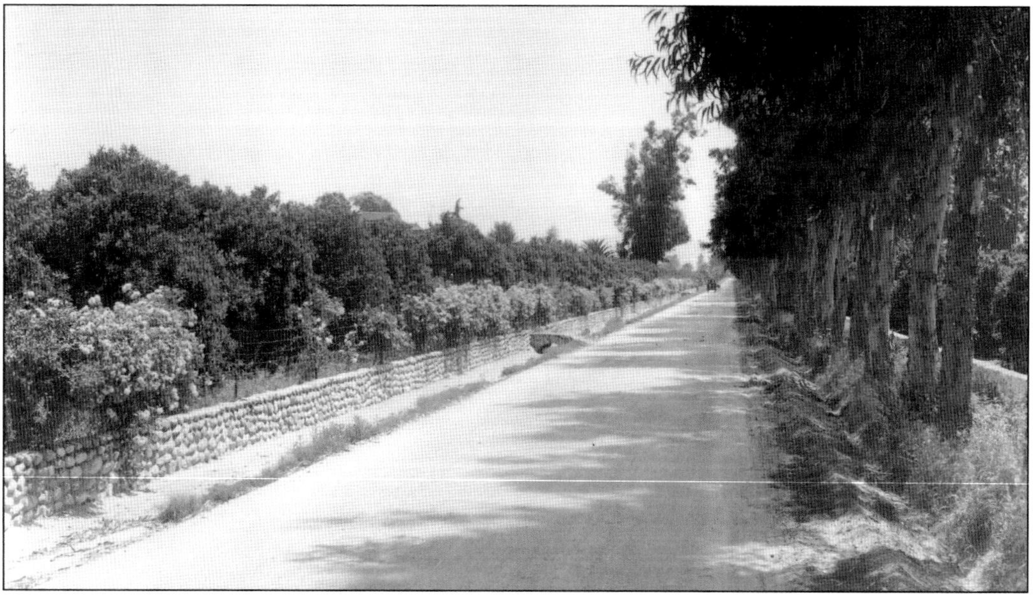

Looking down Citrus Avenue from Gladstone Avenue, it is easy to imagine the Amphion Musical Society members in full evening dress driving their buggies through the citrus groves to attend meetings in Azusa, Covina, and Glendora. The society held meetings on the first Saturday evening after the full moon. Members presented the program. Finding enough hitching places for the horses and keeping the oil lamps burning were the main challenges for the hostess.

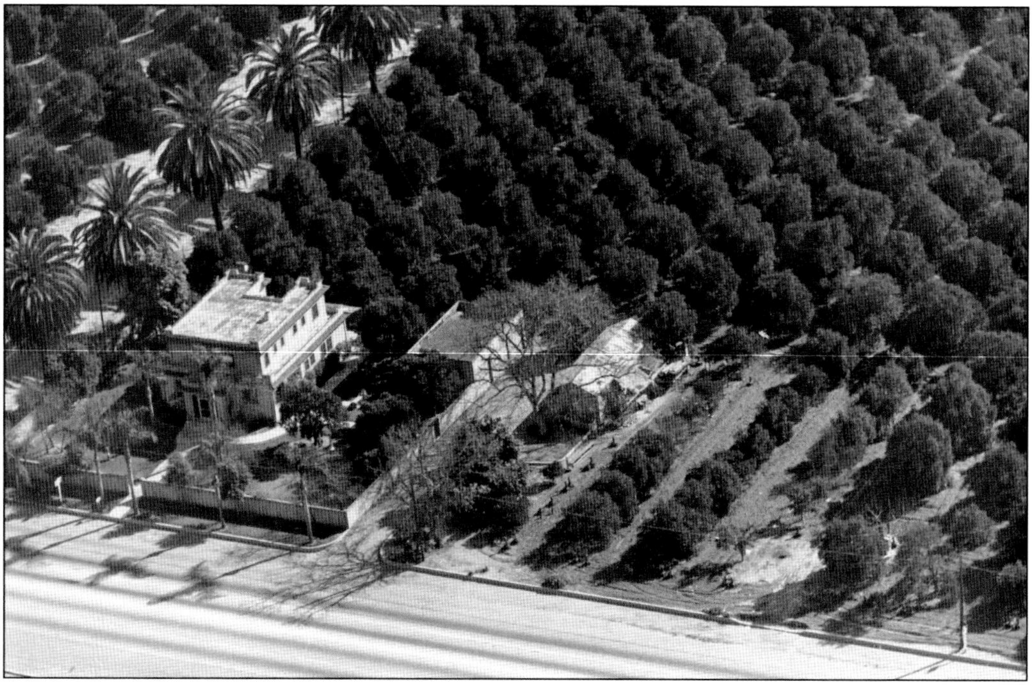

The W.L. Griffith home on southwest Azusa Avenue and East Arroyo (Garvey) Avenue in present-day West Covina was a typical grove house set on a large citrus acreage and surrounded by a walled garden with several varieties of palms. Note the orchard heaters among the trees. Griffith managed the Canfield properties.

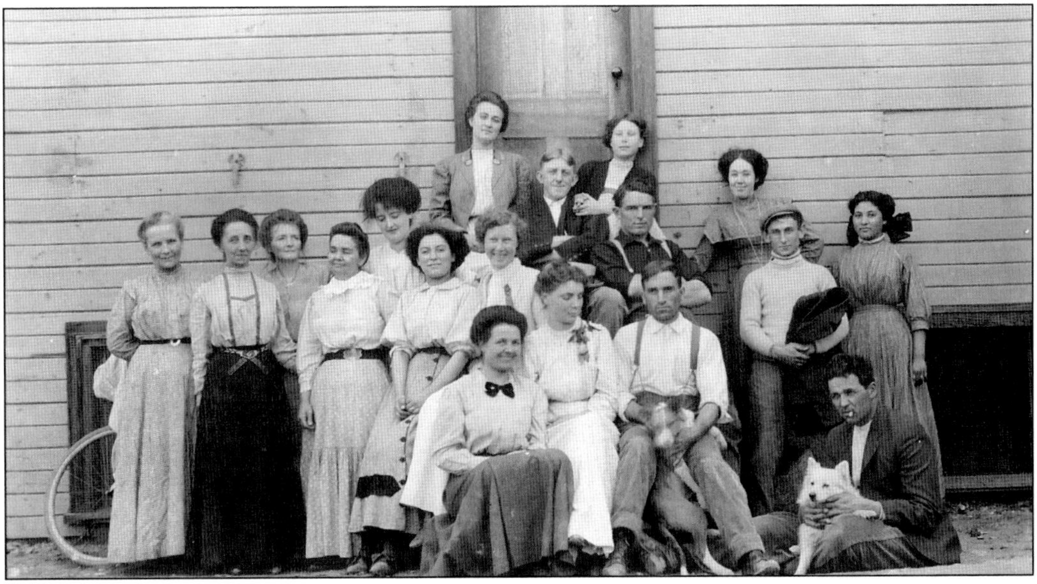

Mailed from Mexico City to Mr. and Mrs. Percy Jackson, of Azusa, California, this postcard is printed in English. The message read, "a Happy Prosperous 1924 be Yours." The Jacksons appear in the ACG packinghouse pictures on pages 28 and 29. (Courtesy of Claire and Bill Stone.)

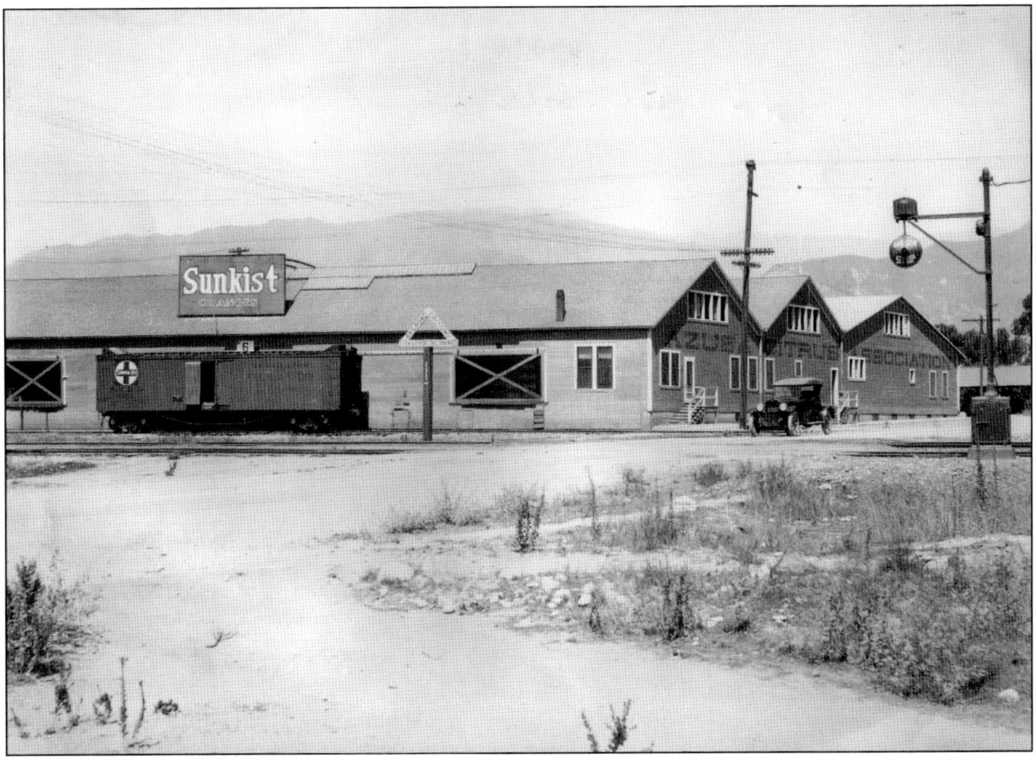

The Azusa Citrus Company was a member of the Sunkist Exchange. It was located north of the Santa Fe Railroad tracks and Alameda Avenue. Operations ceased in the 1950s, and shortly after, the packinghouse was torn down. (Courtesy of Jeffrey Lawrence Cornejo Jr. and the Covina Valley Historical Society.)

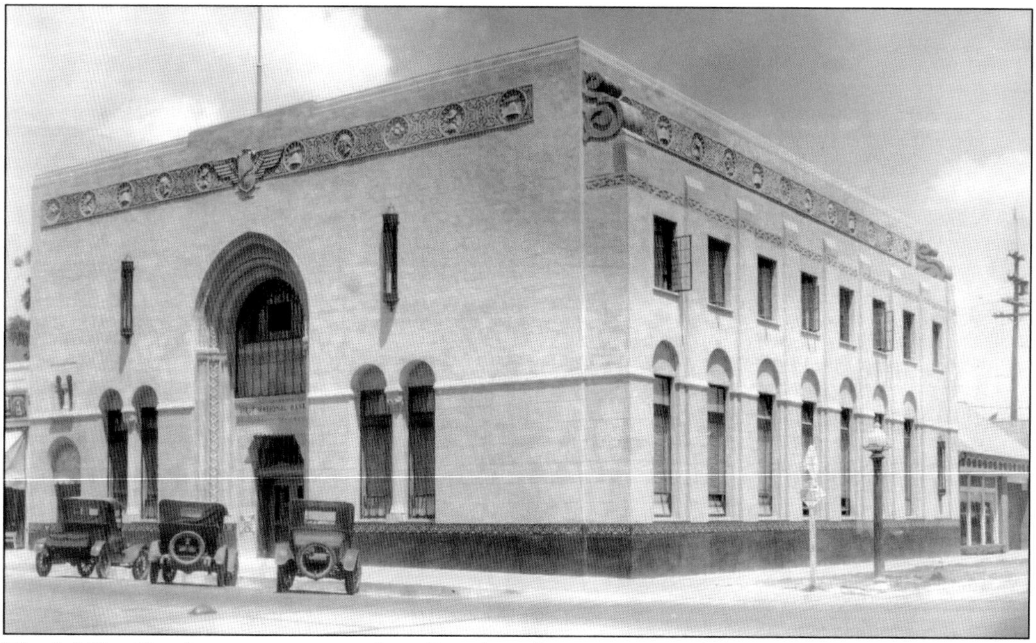

The McKee Brothers, contractors from Los Angeles, constructed the First National and Azusa Valley Savings Bank in 1928. It was designed by architect Robert H. Orr. This two-story, Spanish Moorish–style building is located at 700 North Azusa Avenue. The second floor housed the ACG Fruit Exchange, the city attorneys, the YMCA offices, and the office of author Alec Watkins. (Courtesy of Jeffrey Lawrence Cornejo Jr.)

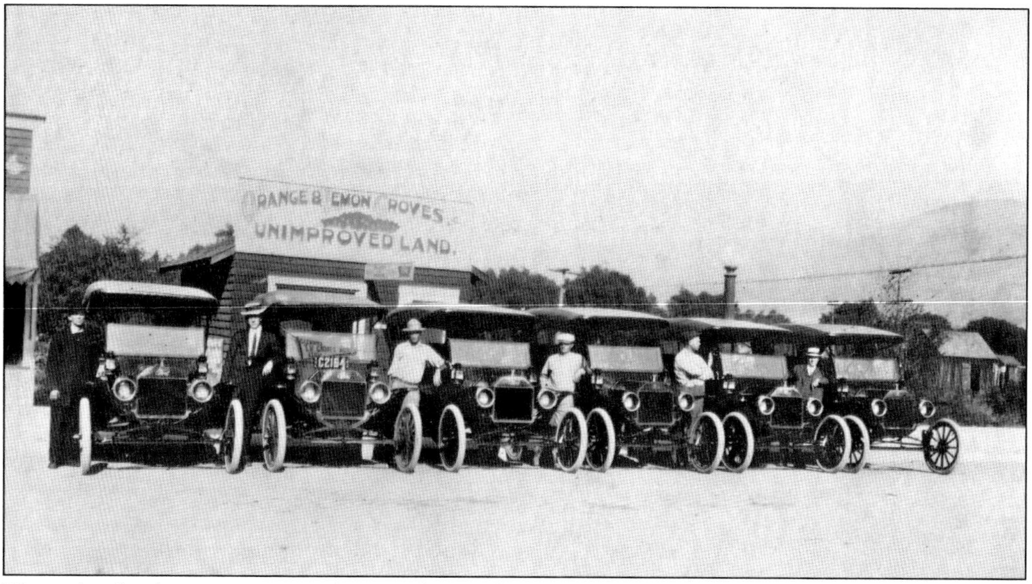

These land salesmen are standing in front of their cars along Foothill Boulevard waiting for customers. The Southern California land boom continued through the 1920s. Failed valley boom towns included Major Gard's Alosta with its elegant hotel, liquor, gambling, and shady ladies; Henry Dalton's Mound City; and Benton. Gladstone, south of Alosta, was named for William Ewart Gladstone, the famous British prime minister who, according to promoters, planned to retire there. (Courtesy of Glendora Historical Society.)

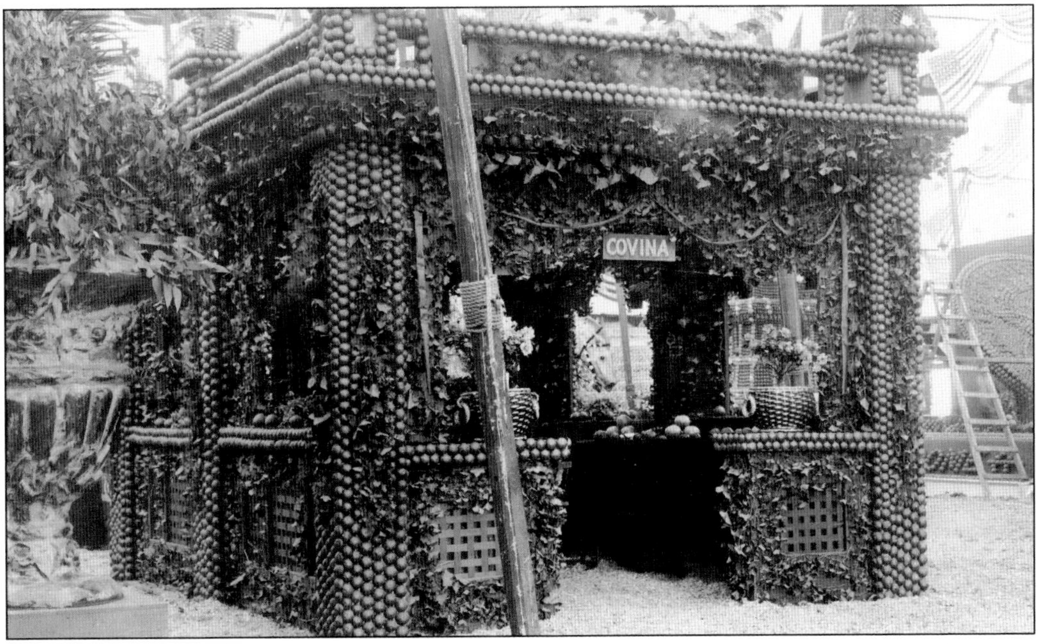

The Renfro Haven Company published this postcard, which features a photograph of a citrus fair. Valley communities sometimes worked together to develop fair displays, including the year when Azusa, Covina, and Glendora created a citrus tower as the centerpiece for a fair. (Courtesy of Claire and Bill Stone.)

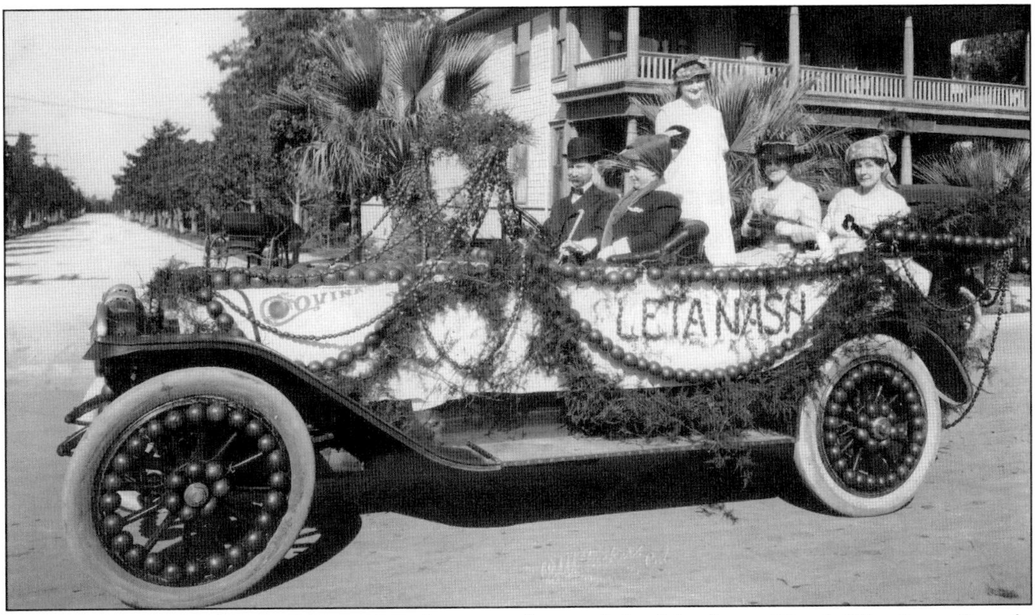

Covina's entry in the 1916 Rose Parade sits in front of the Vendome Hotel. The Rose Parade did not have a queen between 1915 and 1922. Local beauties like Leta Nash, a talented young singer, rode in specially decorated automobiles. William Griswold, manager of the Covina Citrus Exchange, and wife Ava are riding with Nash in the front seat of the Cadillac. Seated in the back are her mother, Sylvia Nash (left), and Delia Matthews. Glendora also entered in the Rose Parade several times.

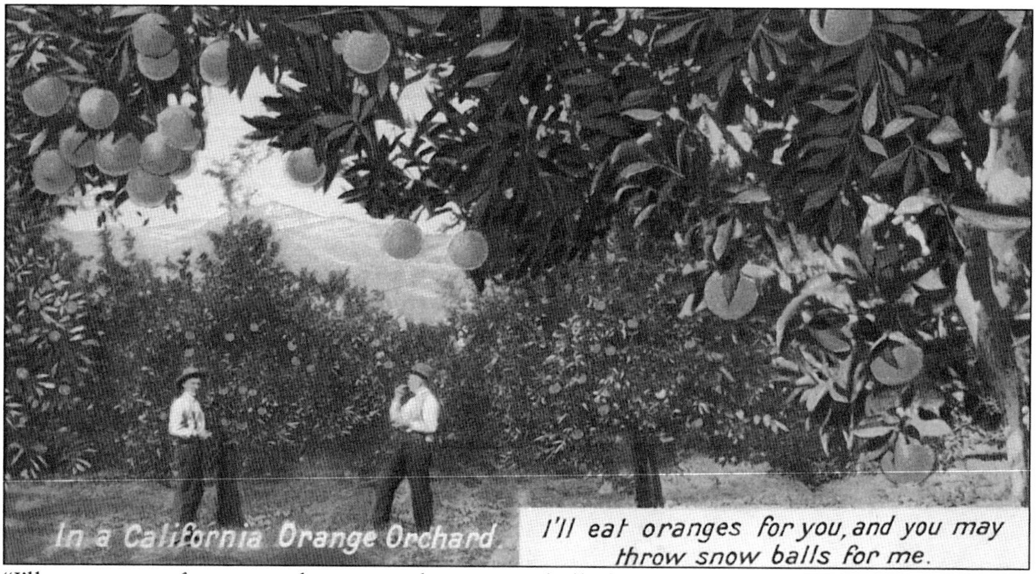

"I'll eat oranges for you, and you may throw snow balls for me" was a very popular postcard for visitors to send back home, particularly during the winter. (Courtesy of Claire and Bill Stone.)

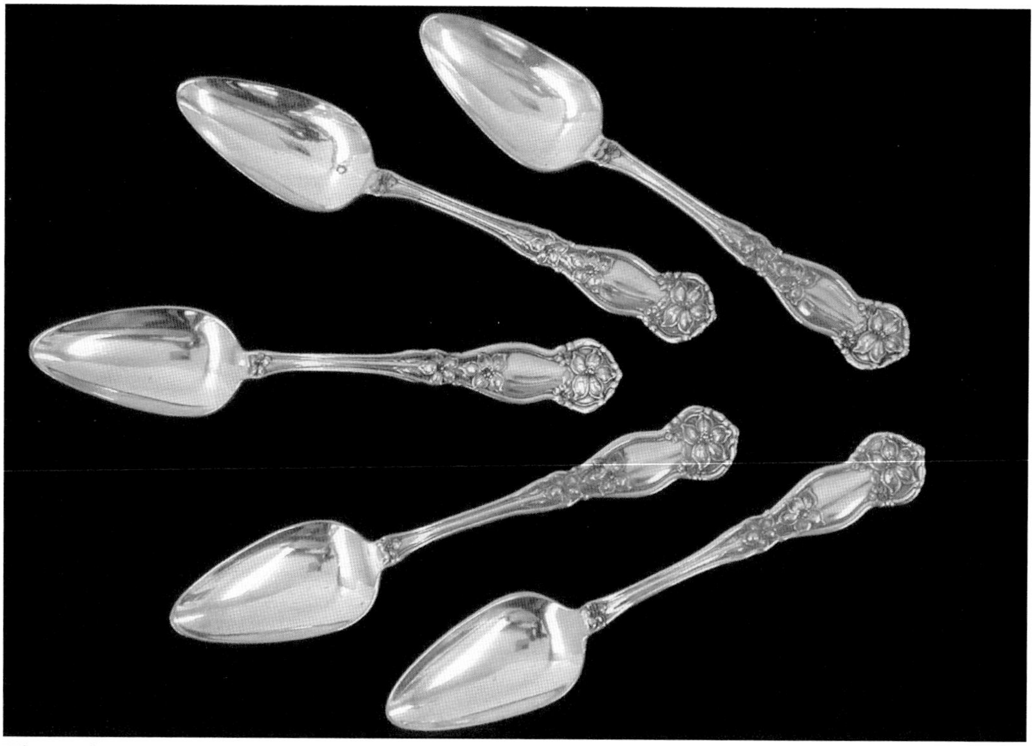
These silver orange spoons in the orange blossom pattern were manufactured by William Rogers and Son, trademark for the International Silver Company. They were offered to retail shoppers who purchased Sunkist products. This was the first retail citrus promotion. Thousands of spoons were redeemed. Citrus crate labels were designed to attract wholesale buyers in the large auction warehouses. Labeled crates staked together made a very striking visual presentation. (Courtesy of Vromans Sliver Shop, Glendora.)

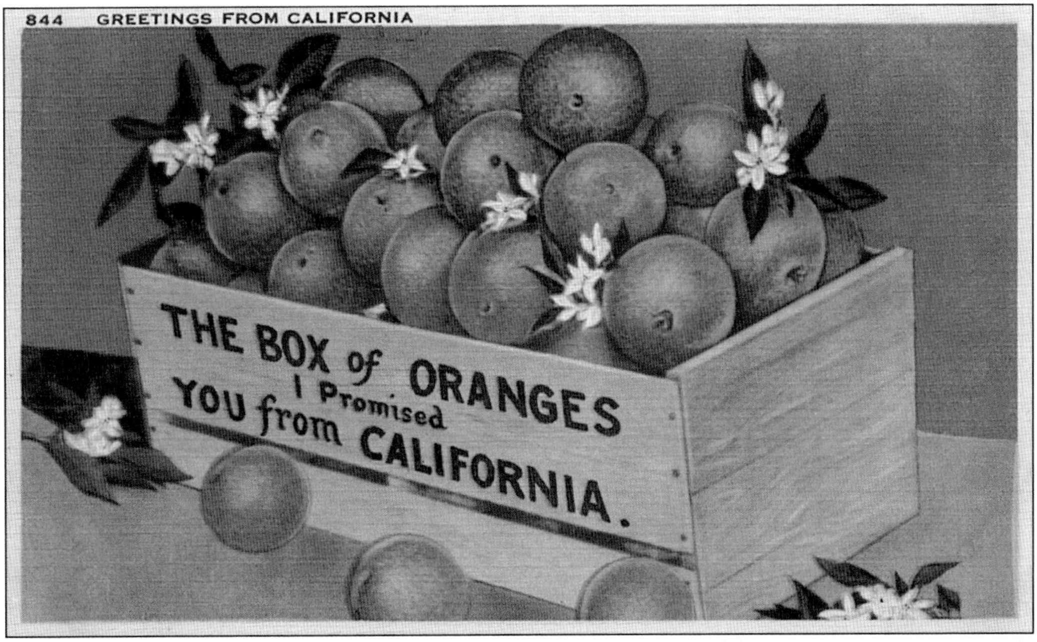

The citrus industry provided many themes for postcards designed for tourists. Some postcards were labeled Sunkist, and some had citrus labels from specific packinghouses. However, most of them—like the one pictured—were generic and could be sold anywhere in California. The earliest postcards were hand-tinted, black-and-white photographs. Later ones were printed in color in Germany. (Courtesy of Claire and Bill Stone.)

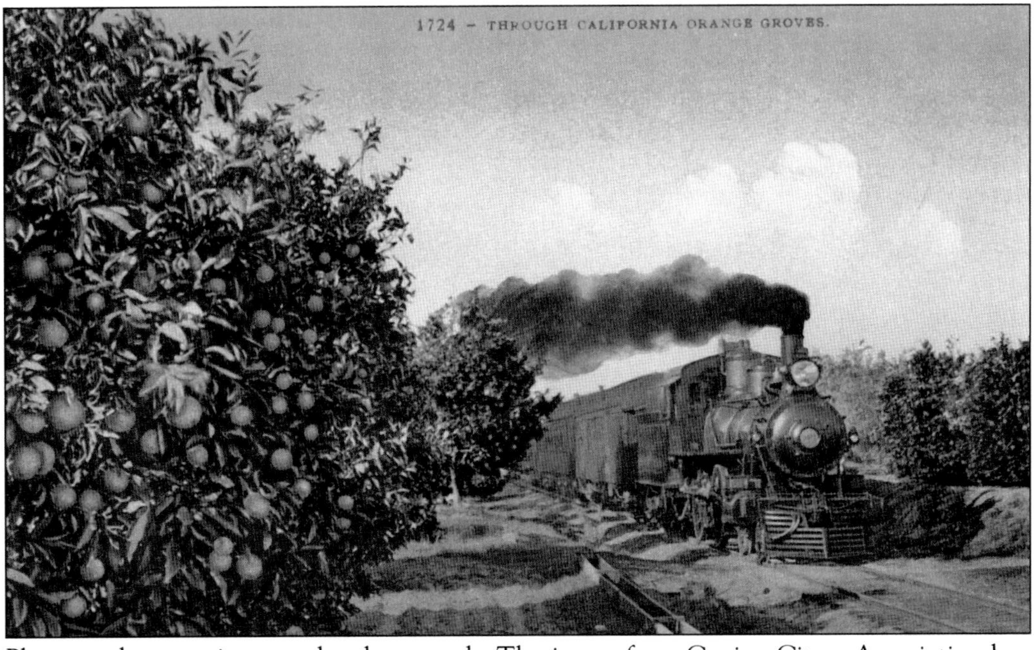

Photographs sometimes predated postcards. The image from Covina Citrus Association box makers from 1898 (shown on page 44) later became a color-tinted postcard design. In the Covina photography studio of C.W. Tucker, who took many of the photographs in this book, Lucy Wheeler and Tucker's wife, Grace, tinted the postcards by hand. (Courtesy of Marty Getz.)

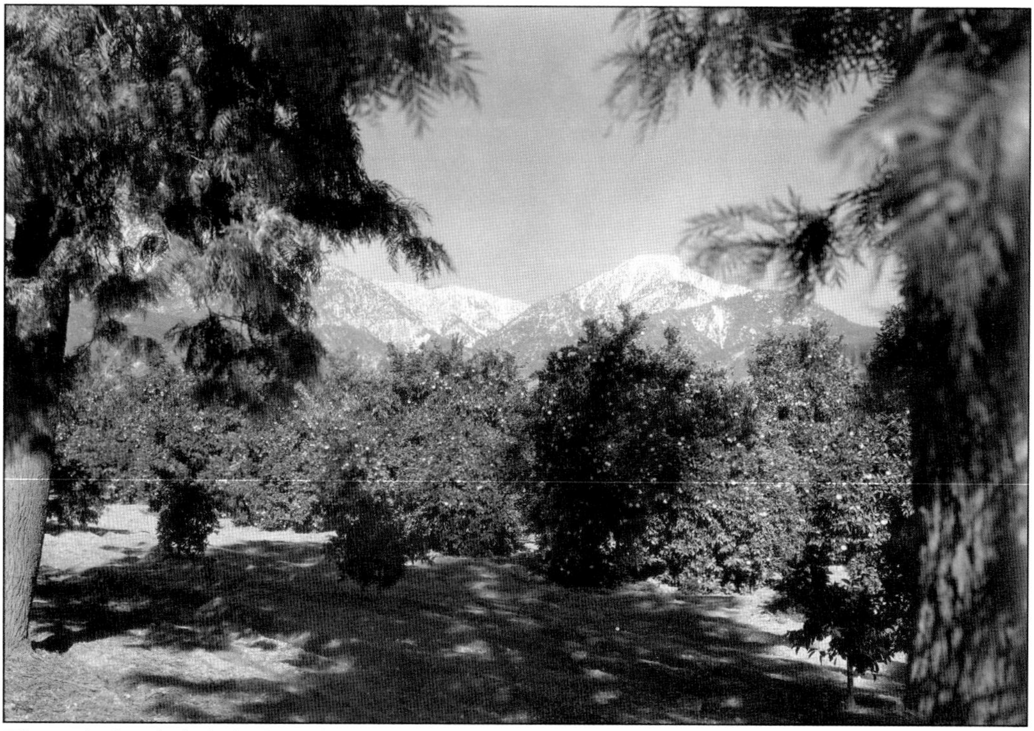

The soil, the slightly higher elevation, and the frost-free climate of Glendora and Charter Oak made these areas the best locations in the valley to grow lemons. This Sunkist photograph shows a lemon grove in the Glendora foothills with snow-covered Mount San Antonio in the background. (Courtesy of Sunkist.)

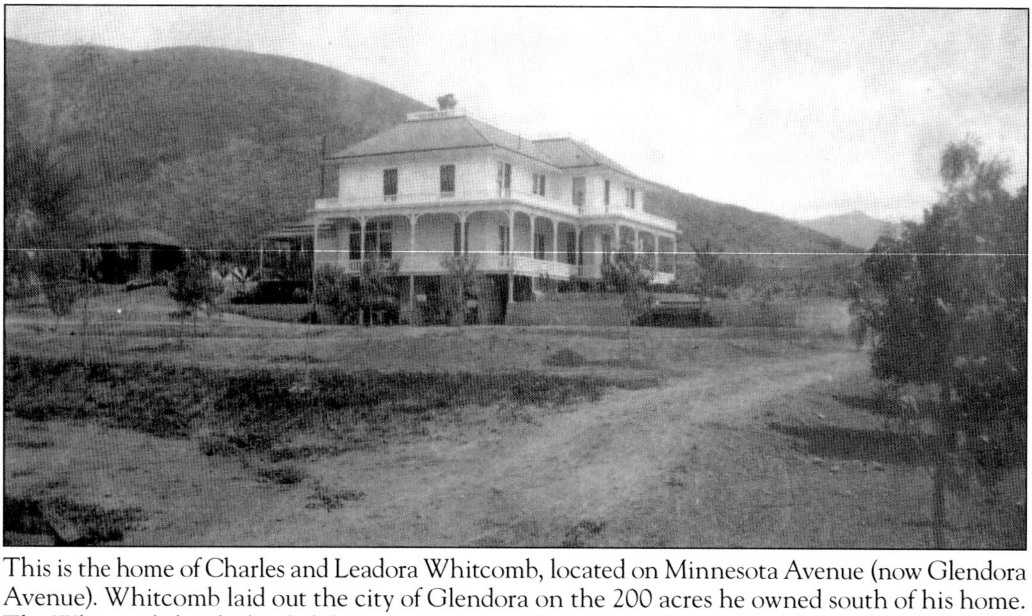

This is the home of Charles and Leadora Whitcomb, located on Minnesota Avenue (now Glendora Avenue). Whitcomb laid out the city of Glendora on the 200 acres he owned south of his home. The Whitcomb family divided their time between Glendora and Chicago, where they had business interests. When Charles and Leadora moved to Santa Monica, one of their daughters opened a school at the house. (Courtesy of Glendora Historical Society.)

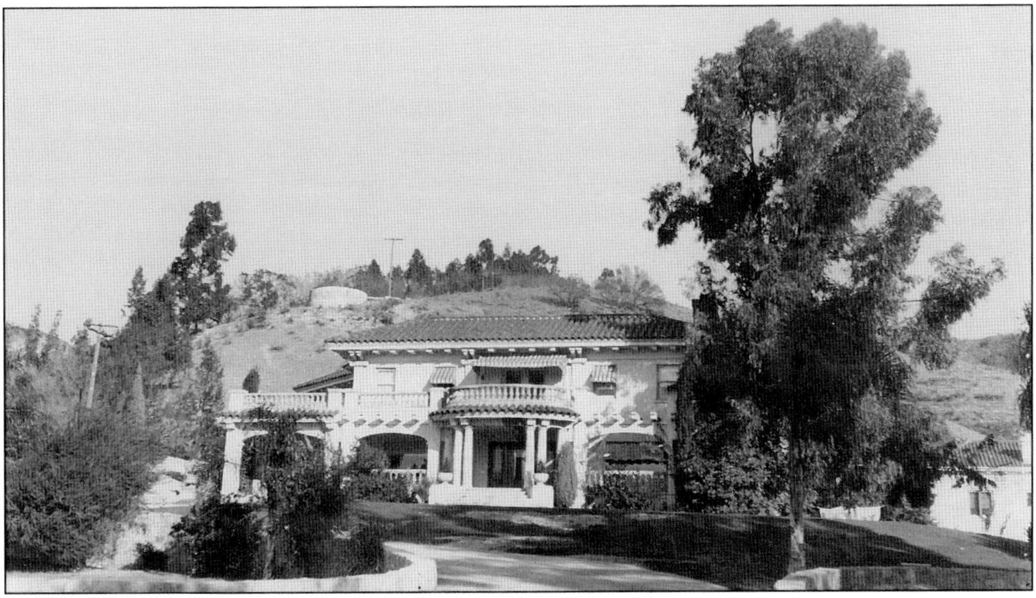

In 1898, pioneer citrus rancher Charles Warren purchased land in Glendora east of Amelia and north of Alosta. His two-story Italian villa was built in 1915. Archer was the architect, and George Fickelt was the contractor. At the time, it was the most elegant house built in the valley. It had tile and hardwood floors and art-glass doors. The exterior featured a man-made lake and terraced flower gardens. (Courtesy of Glendora Historical Society.)

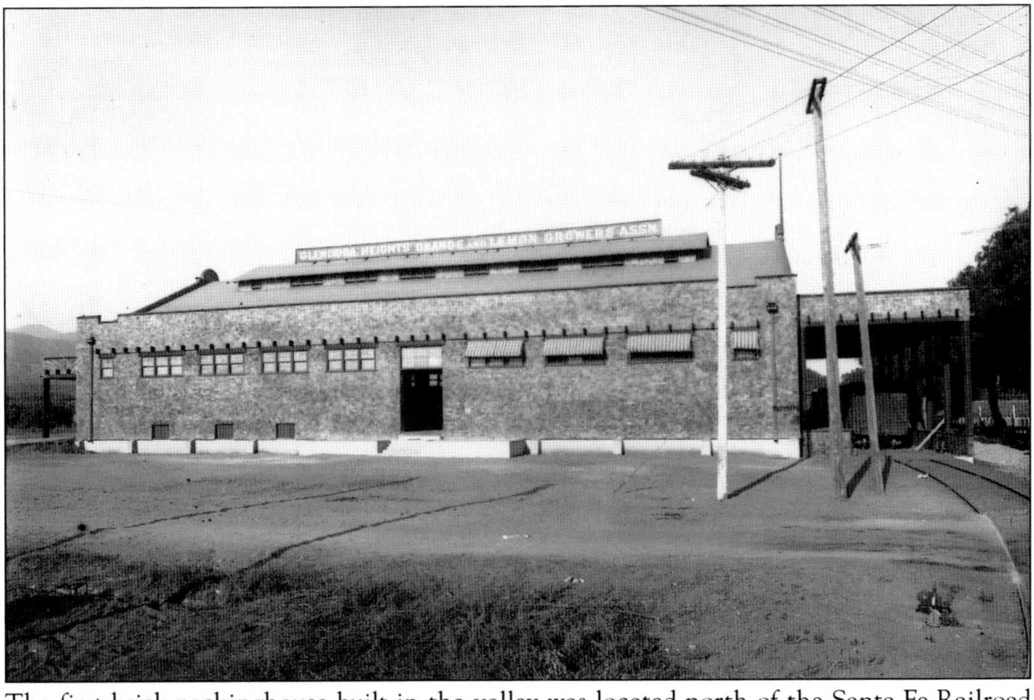

The first brick packinghouse built in the valley was located north of the Santa Fe Railroad tracks in Glendora. Glendora oranges were favored by Pres. William Howard Taft. During his administration, a monthly shipment of Glendora oranges was sent for the White House table. (Courtesy of Glendora Historical Society.)

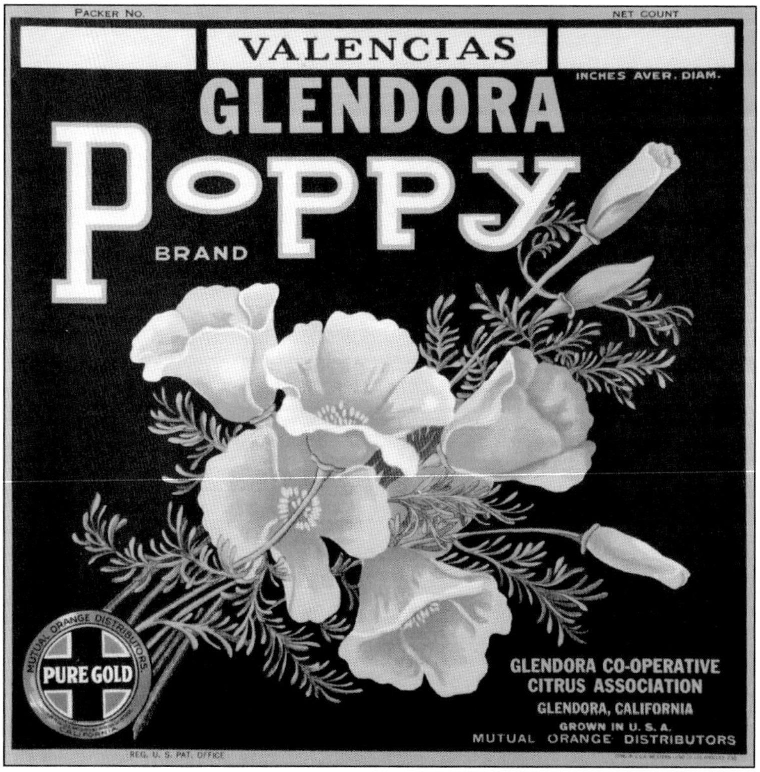

The Glendora Valencia Poppy label was printed by Los Angeles Western Lithograph Company for the independent Glendora Cooperative Citrus Association. Mutual Orange Distributors was the shipping agent.

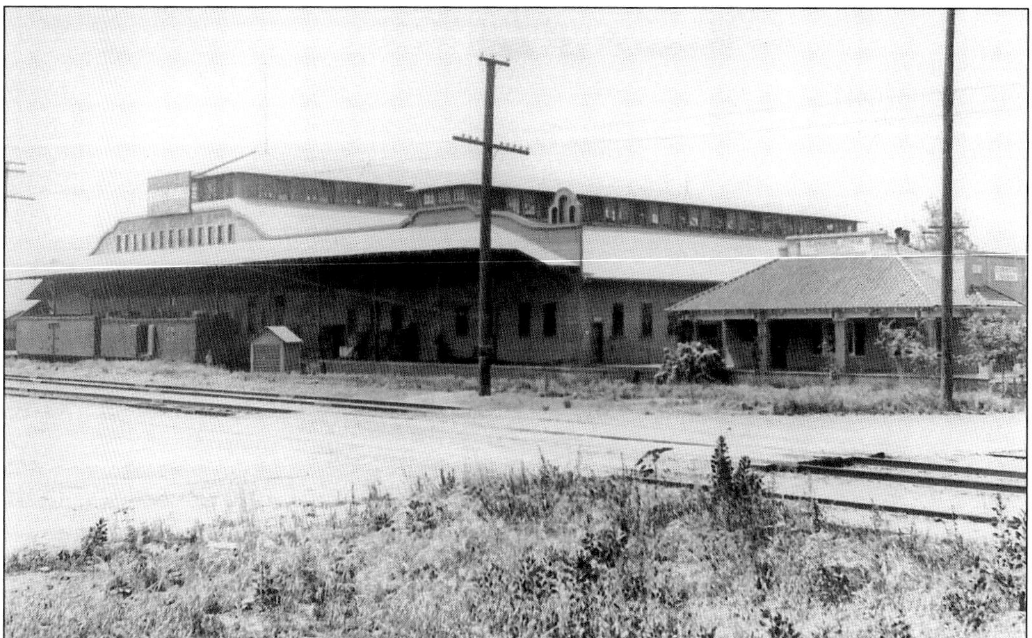

The first Glendora Citrus Association packinghouse at Ada and Vermont in Glendora was located on the north side of the Santa Fe Railroad tracks. (Courtesy of Glendora Historical Society.)

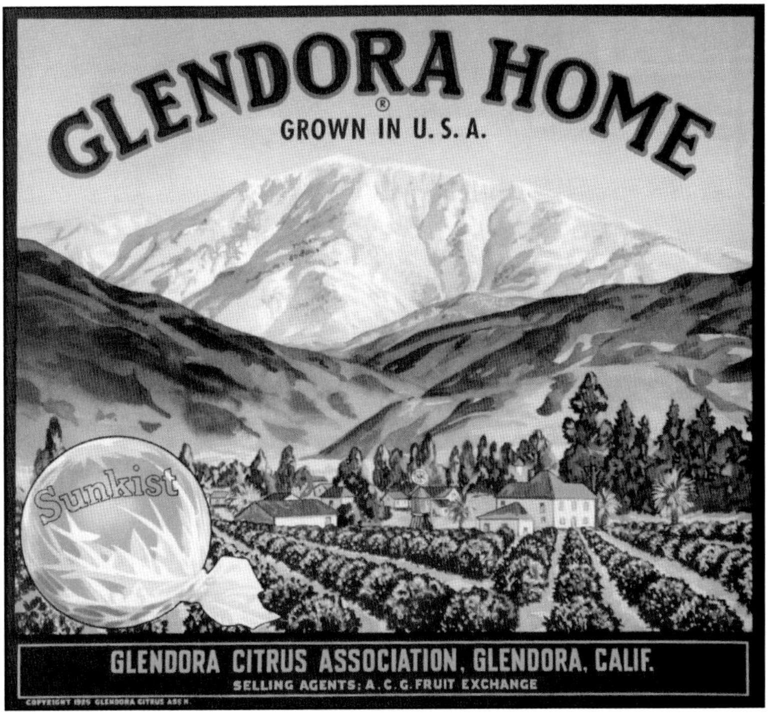

Glendora Home, the most famous valley citrus label, was designed and printed by the Carton Label Corporation for the Glendora Citrus Association. The ACG Fruit Exchange was the shipping agent. It was the last Glendora Citrus Association label shipped by the ACG. Glendora started its own Fruit Exchange in 1923. The image of the snow-covered Mount San Antonio looming over a settlement of citrus groves became a symbol of the valley.

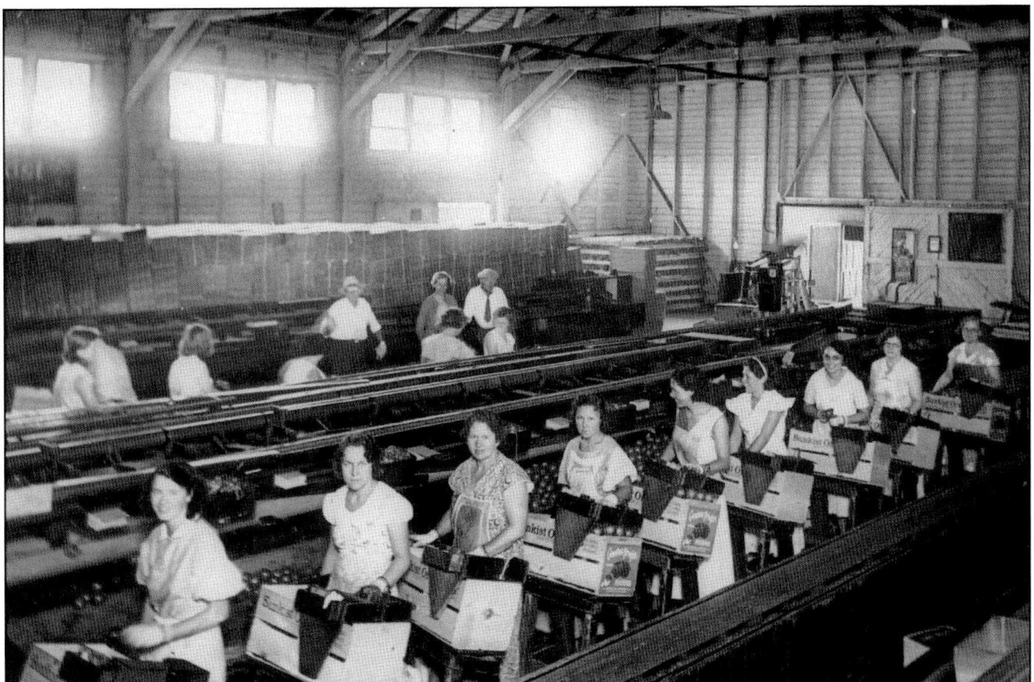

This is an interior photograph of the Glendora Citrus Association packinghouse at Ada and Vermont Avenues. Notice the modern packing equipment and the changes in the women's hairstyles and clothes. (Courtesy of Glendora Historical Society.)

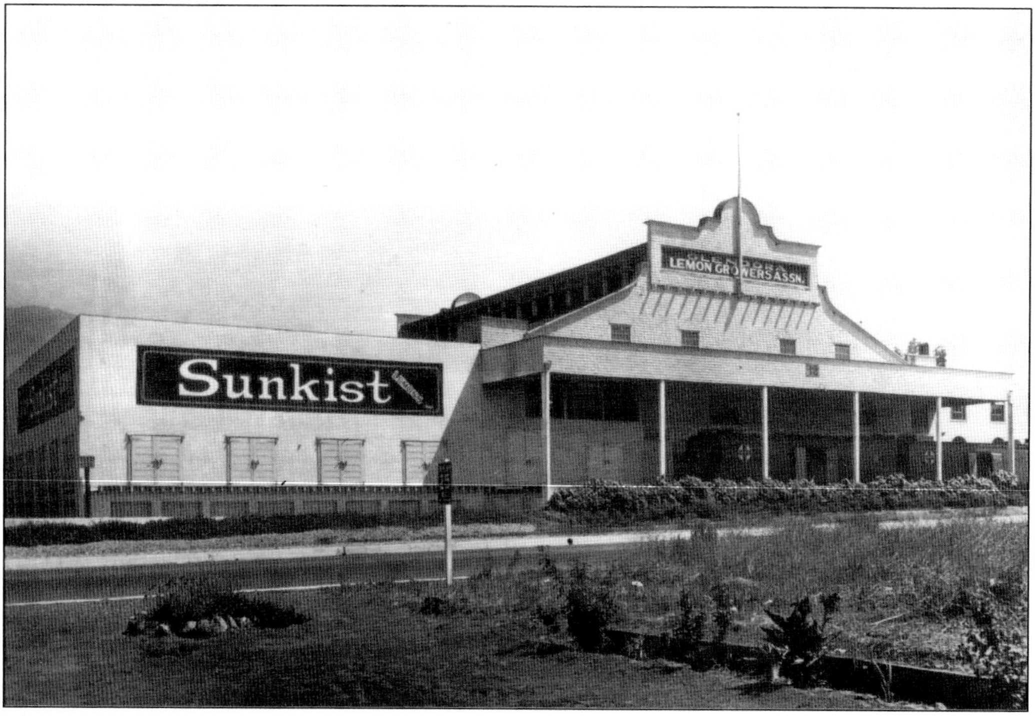

This new Sunkist packinghouse was located in Glendora by the Santa Fe Railroad tracks. It was built in the 1920s. By 1940, there were five orange packinghouses and one lemon packinghouse in Glendora. The photograph above shows an exterior view of the facility. In the photograph below, the interior is shown with workers. The packinghouse was outfitted with the latest crate-making machinery. (Both, courtesy of Glendora Historical Society.)

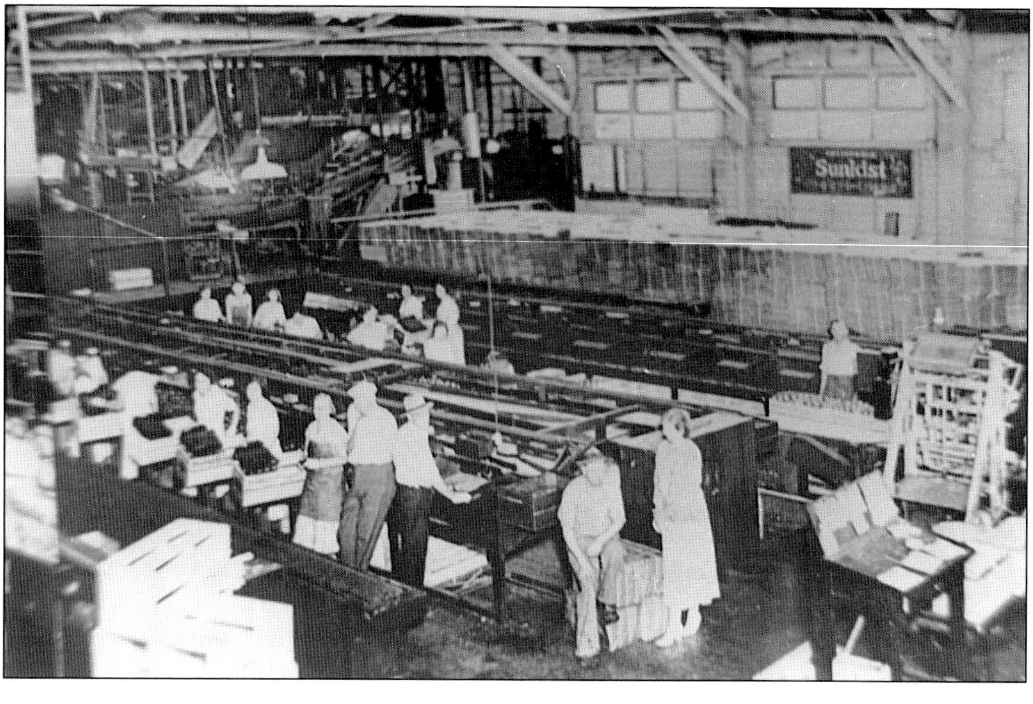

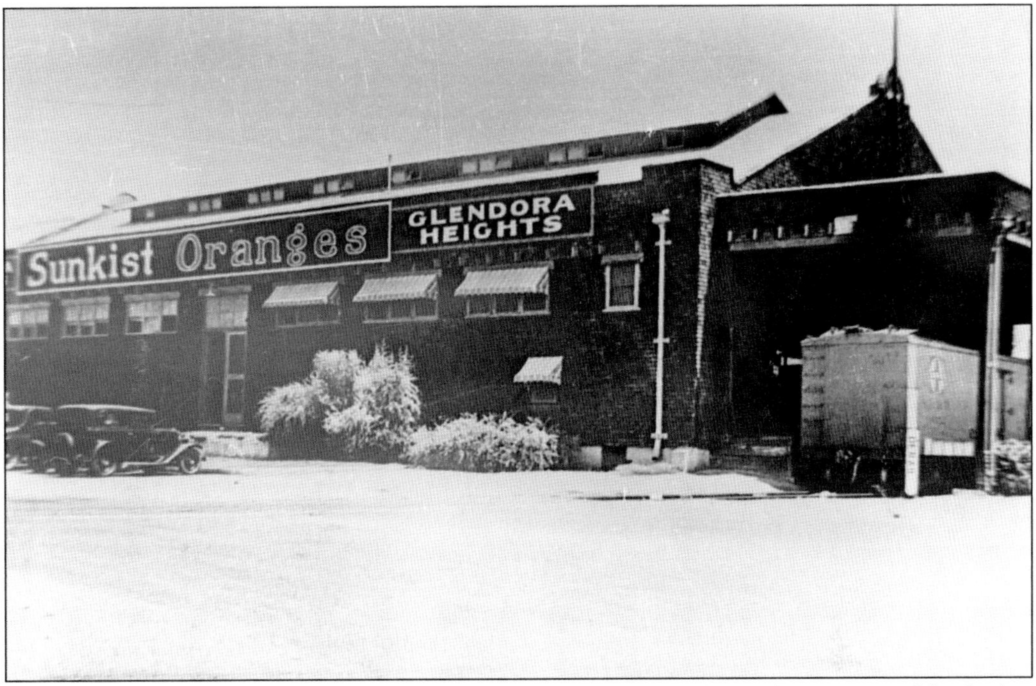

The Glendora Heights Grange and Lemon Association was formed in 1910 and built this packinghouse on the northeast corner of Grand and Foothill Boulevard. (Courtesy of Glendora Historical Society.)

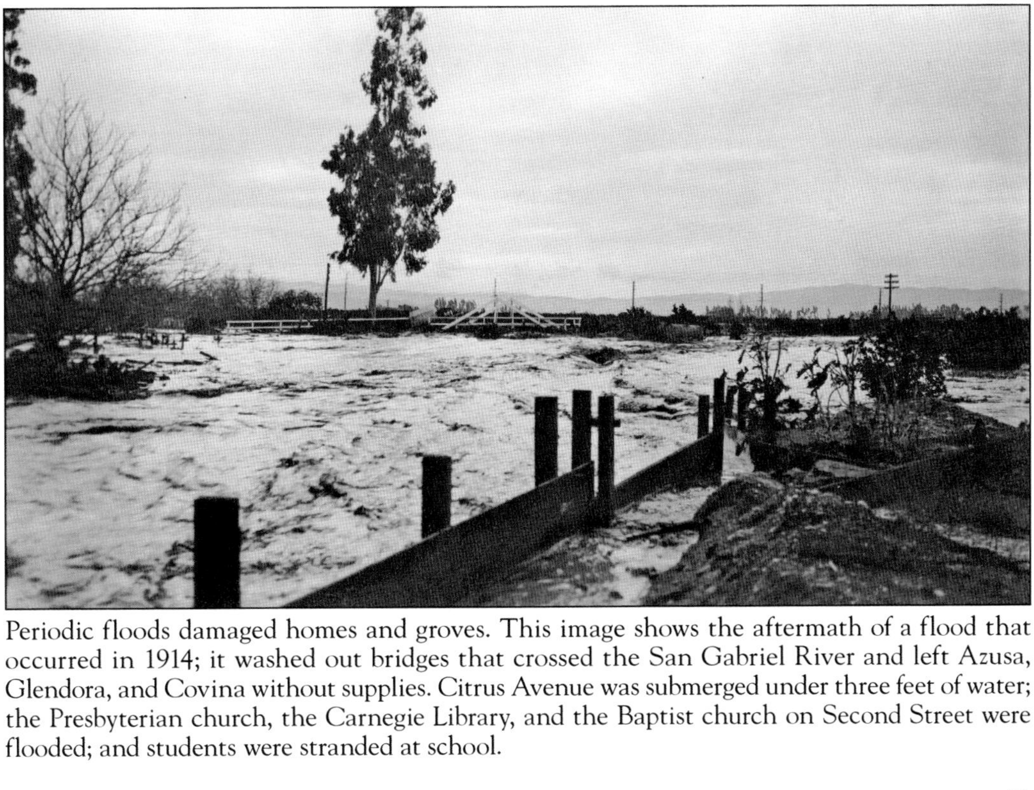

Periodic floods damaged homes and groves. This image shows the aftermath of a flood that occurred in 1914; it washed out bridges that crossed the San Gabriel River and left Azusa, Glendora, and Covina without supplies. Citrus Avenue was submerged under three feet of water; the Presbyterian church, the Carnegie Library, and the Baptist church on Second Street were flooded; and students were stranded at school.

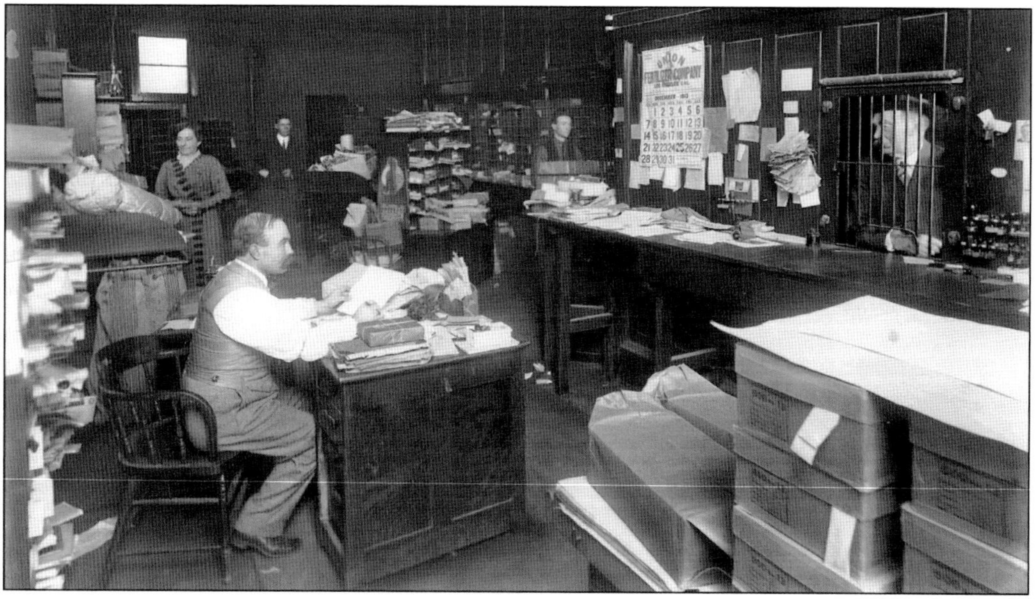

James Louis Matthews owned and published the *Covina Argus*. His newspaper chronicled the development of the valley citrus industry from its earliest days. He directed two Los Angeles County bond drives, raising $35 million for the construction of 13 dams in the San Gabriel Mountains. Matthews also was the first president of the Covina Rotary, and he served as postmaster from 1904 to 1916.

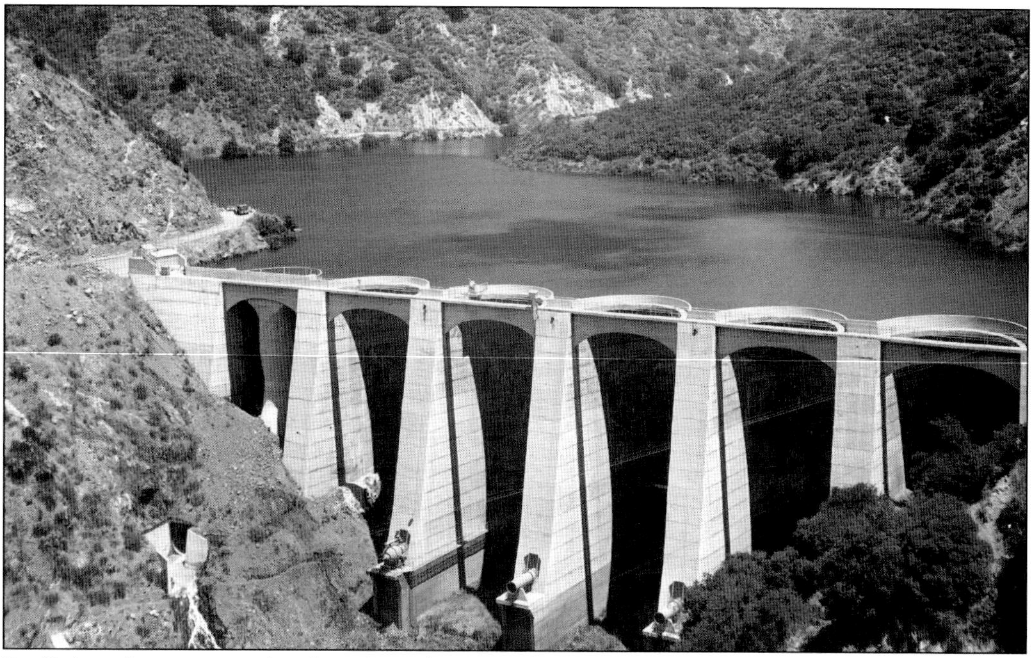

Big Dalton dam was completed in 1929. The Big Dalton wash runs from the San Gabriel Mountains to the South Hills, which are one and a quarter miles east of Glendora. The wash turns northward just south of the south hills. Before the dam was completed, the water level would rise dramatically during the rainy season. The dam's capacity is 1,200 acre-feet of water, and the cost was $911,000. (Courtesy of Glendora Historical Society.)

A row of eucalyptus trees lines the Rowland Avenue center median just west of Hollenbeck Street. Mature valley citrus groves had reached maximum production. In 1919, Charter Oak Sunkist packinghouse shipped 149,688 boxes of Washington Navel oranges and 325 carloads of Valencias. The independent Covina Heights Groves, Inc., shipped 29,568 boxes of Washington Navels at $3.11 each and 50 carloads of Valencias at the same price.

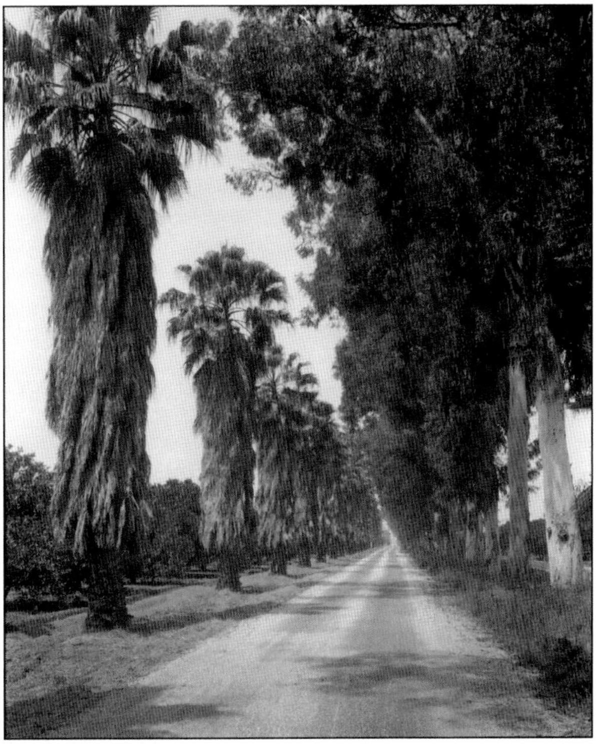

This is the rose-covered entrance to the beautiful, Craftsman-style home and citrus grove of Dr. Claude Searcy, a dentist who practiced in Covina. The house was on Covina Boulevard east of Grand Avenue. Unlike the early days when valley residents hired Dr. Burdick to take care of them, the citrus communities now each had their own doctors and dentists.

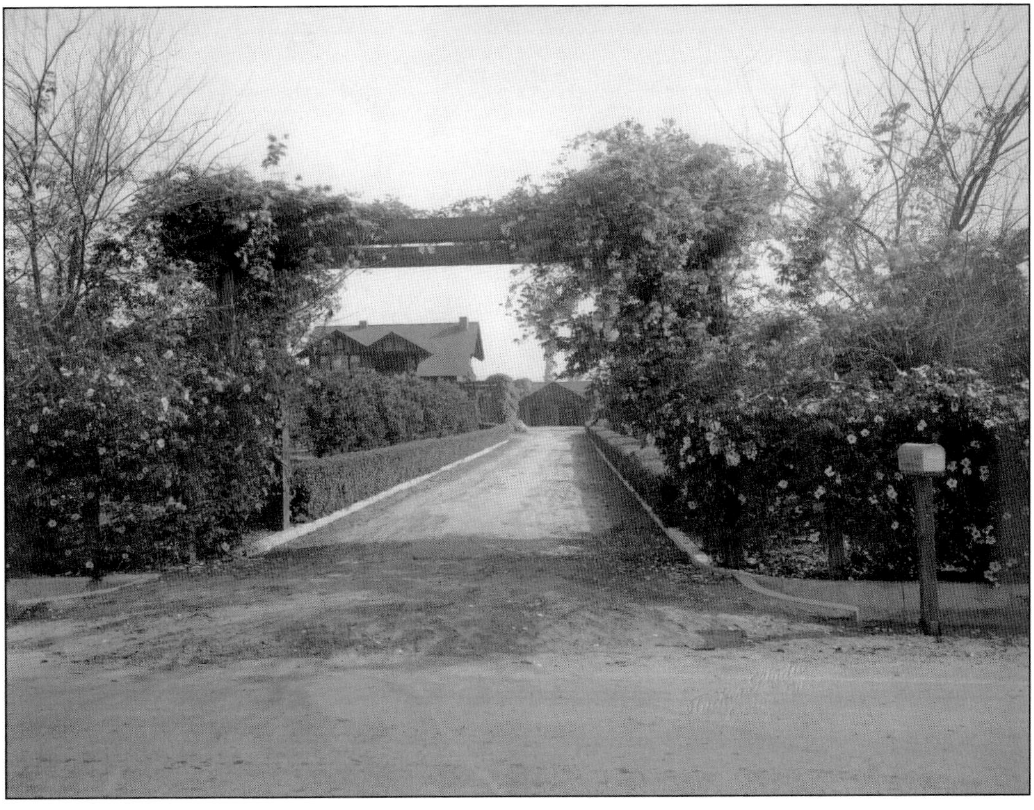

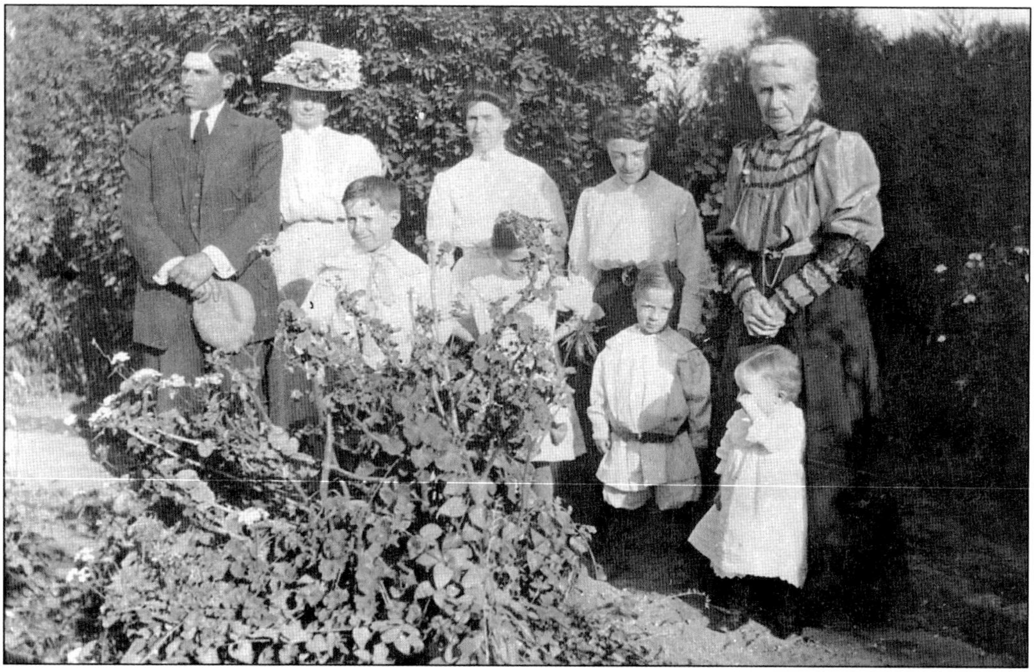

Gertrude Damerel assisted Mary Leech, her widowed sister-in-law, with raising her children. Pictured are, from left to right, (first row) Edward, Elea, William, and Fanny Leech; (second row) Harry and Gertrude Damerel, Mary Leech, unidentified, and Sarah Damerel. Harry Damerel helped make Covina the largest orange shipping center in California. With Arden Dairy, he pioneered home delivery of fresh orange juice, and he installed the first equipment to freeze orange juice.

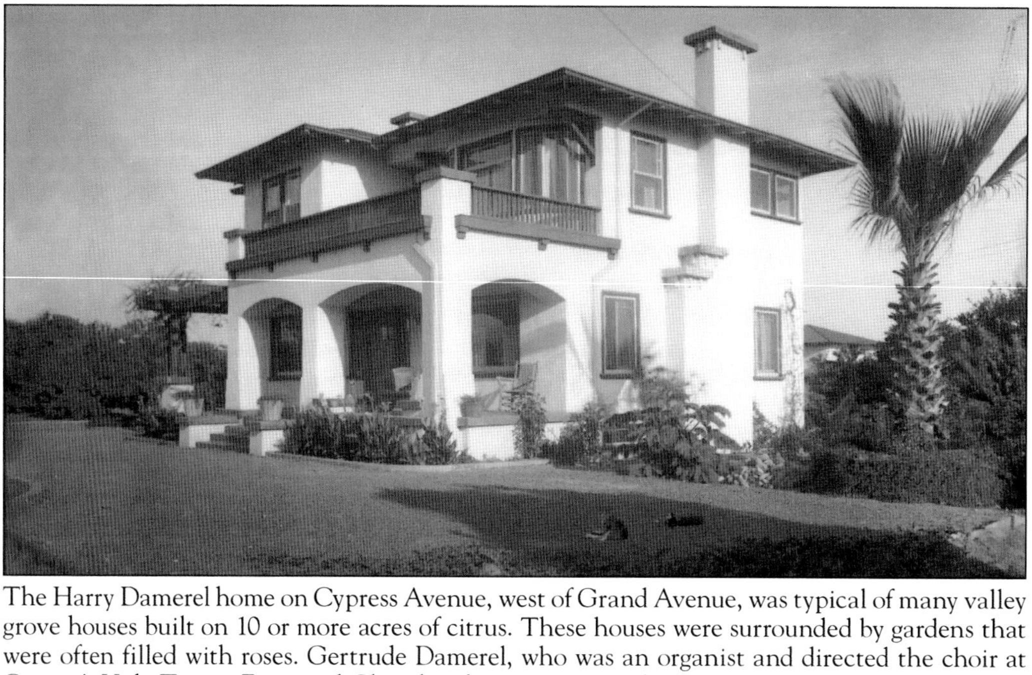

The Harry Damerel home on Cypress Avenue, west of Grand Avenue, was typical of many valley grove houses built on 10 or more acres of citrus. These houses were surrounded by gardens that were often filled with roses. Gertrude Damerel, who was an organist and directed the choir at Covina's Holy Trinity Episcopal Church, often gave musical teas at her home for the benefit of the church.

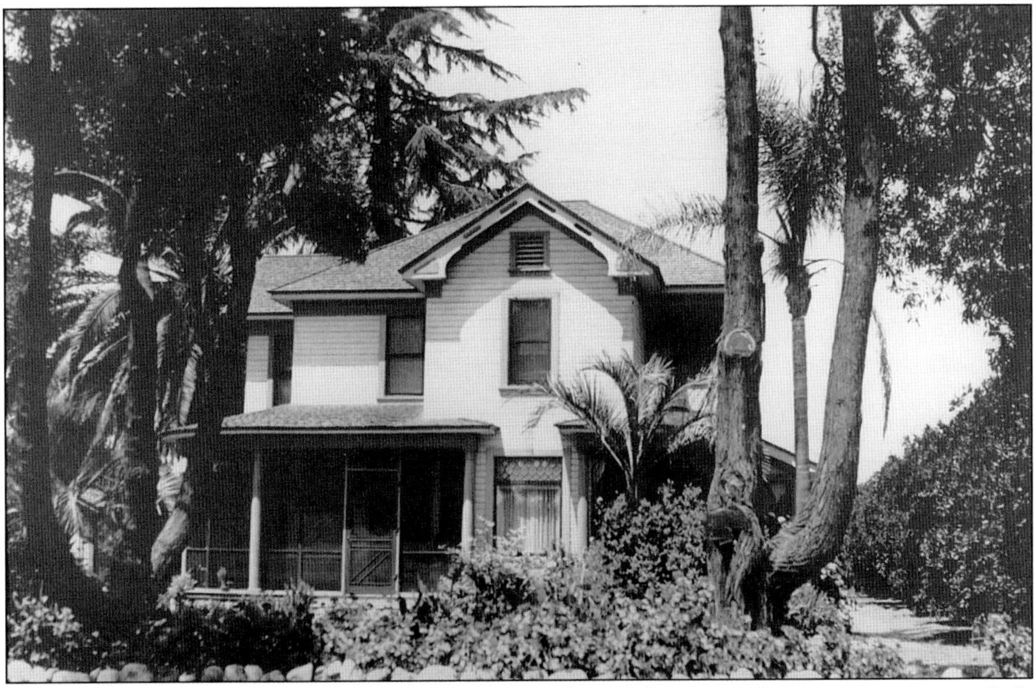

The availability of transportation allowed many professionals to enjoy life in a citrus grove while being employed elsewhere. Oscar Ihsen, an architect, and Herman Ihsen, a metallurgical engineer were typical of this new wave of valley residents. Their home, built in the 1890s, was in a 10-acre orange grove in Charter Oak. After the brothers married, Oscar moved to Covina.

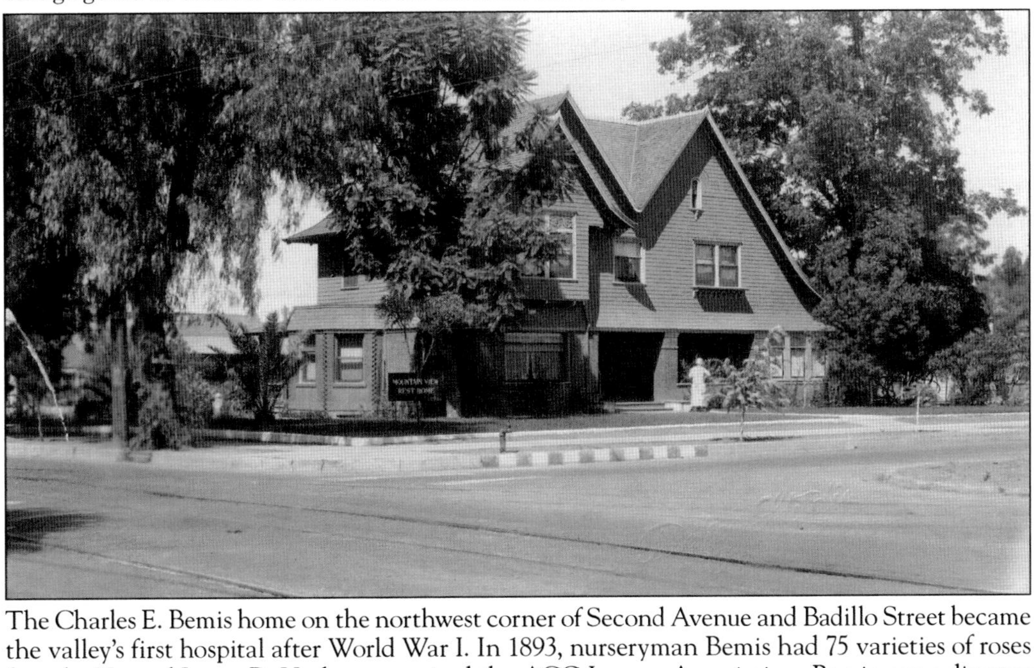

The Charles E. Bemis home on the northwest corner of Second Avenue and Badillo Street became the valley's first hospital after World War I. In 1893, nurseryman Bemis had 75 varieties of roses for sale. He and James R. Hodges organized the ACG Lemon Association. Bemis was a director of the Covina Citrus Association, and in 1900, he became Covina's first justice of the peace and horticultural inspector for the valley.

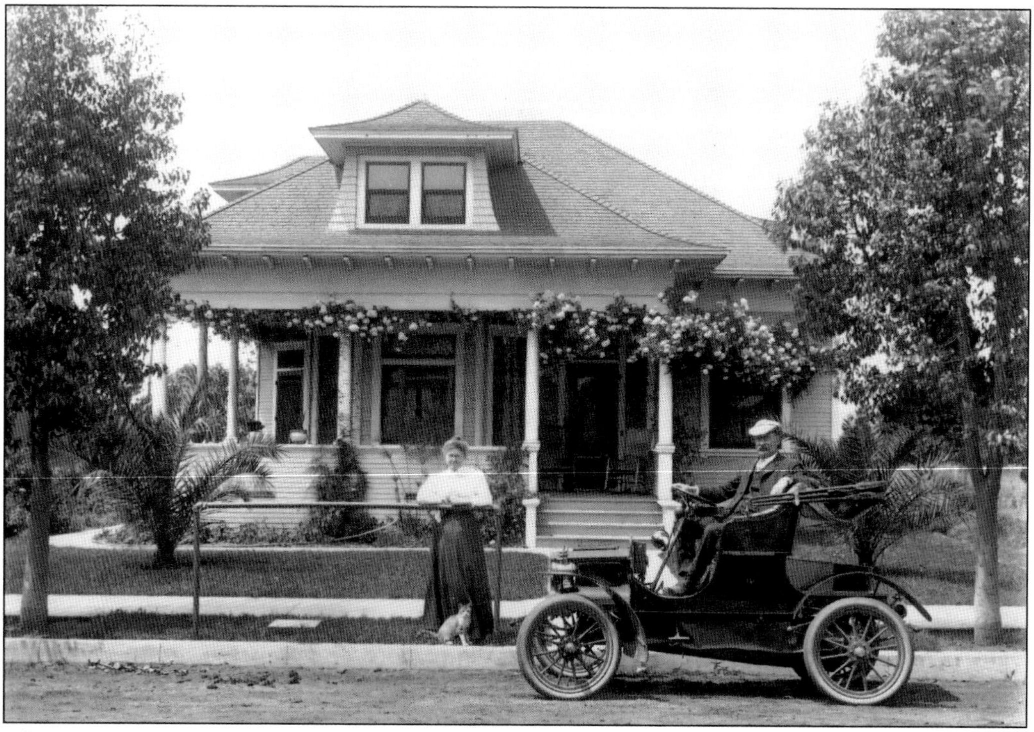

Clara Anderson is standing with the family cat next to a hitching post in front of her California-style bungalow home on San Bernardino Road. George Anderson, manager of the Covina Citrus Association packinghouse, is proudly displaying his new Reo. Roses are blooming on the porch. Bungalow-style homes still exist in Covina on Puente, Dexter, Center, School, Italia, Cottage, Orange, Rowland, and Badillo Streets. There are also many in Azusa and Glendora.

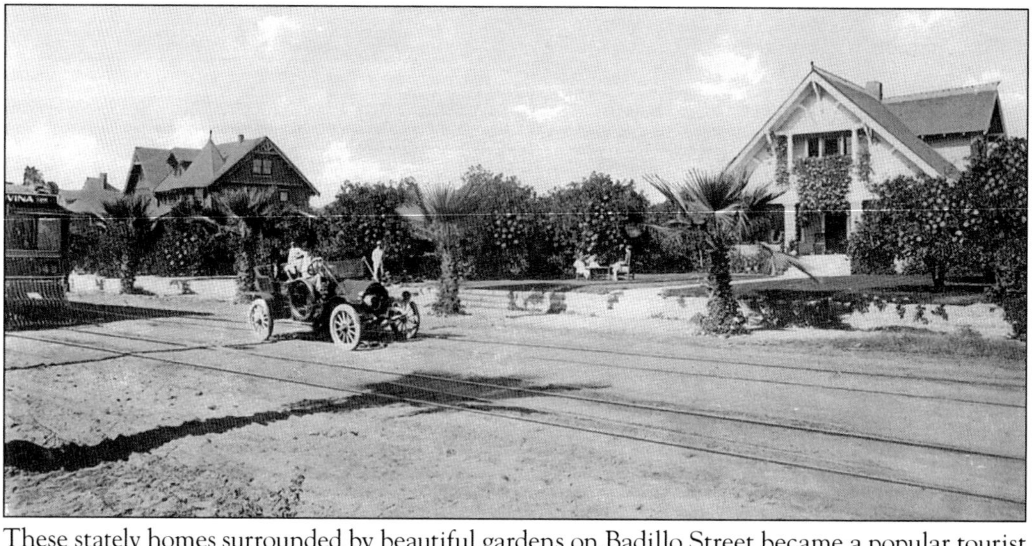

These stately homes surrounded by beautiful gardens on Badillo Street became a popular tourist attraction on the Pacific Electric Orange Empire Trolley trip through the Covina and Redland citrus orchards. Covina's flowers were famous, especially the roses. When Bess Aschenbrenner was president of the Covina Woman's Club, Elgenia Behrons Bixby founded and became the curator of the first garden section formed by a woman's club in the United States.

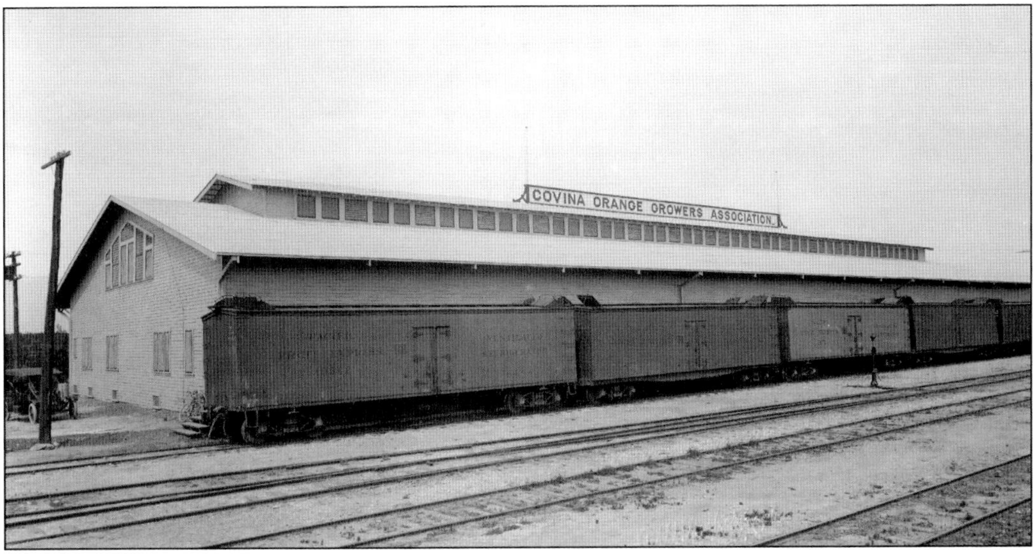

The Covina Orange Growers Association was rebuilt after it burned down on January 2, 1914. Contractor A.W. Feist built the new packinghouse with the latest machinery; it was finished in time for the start of the Valencia season. Citrus packing continued at the facility until 1954 when the packinghouse was sold to the Revolving Fund Investment Company. It was purchased by James Hodges and Jack Cone in 1962 and demolished in 1974.

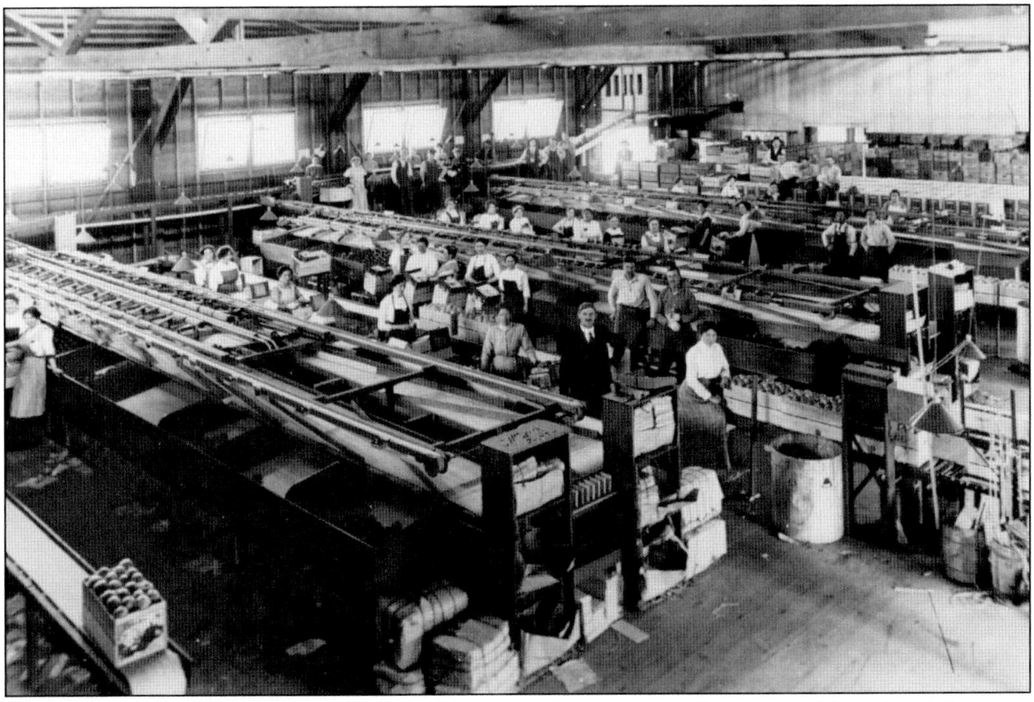

This is the new Orange Growers Association packinghouse interior. Notice the Daisy-labeled orange crates. In 1926, Bill Hamlin invented a creamy orange drink for his friend Julius Freed, who had an orange juice stand in downtown Los Angeles. The drink was an instant success. Many people came by, saying to Freed, "Give me an orange, Julius." The name was born. By 1929, Hamlin had established 100 Orange Julius stores. One of the first was in Covina, Hamlin's hometown.

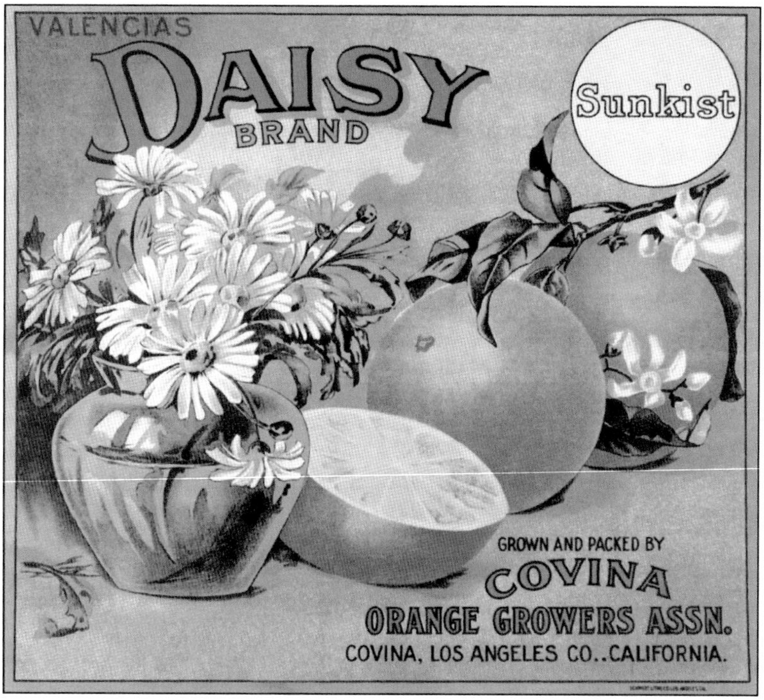

The Daisy Brand Valencia label was designed and printed by Los Angeles Schmidt Lithograph Company for the Covina Orange Growers Association. The Sunkist trademark was printed in a yellow circle. The daisies are in a vase next to a cut orange and two whole Valencias hanging from a branch with somewhat realistic orange blossoms.

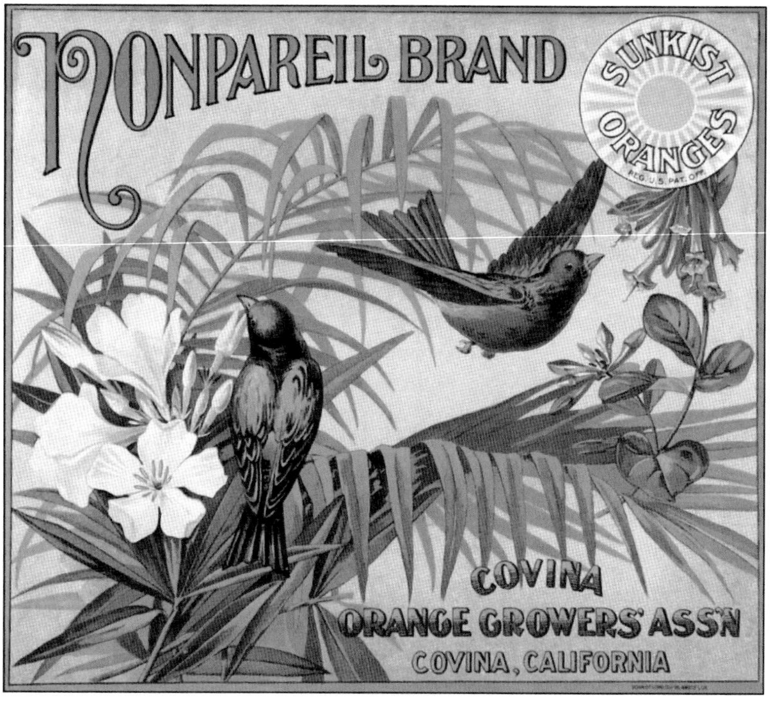

The Nonpareil Brand label was designed and printed by Los Angeles Schmidt Lithograph Company for Covina Orange Growers Association. The word nonpareil means "better than anything else." The label, with its fanciful arrangement of palm leaves, birds, and non-citrus blossoms, carries the Sunkist medallion.

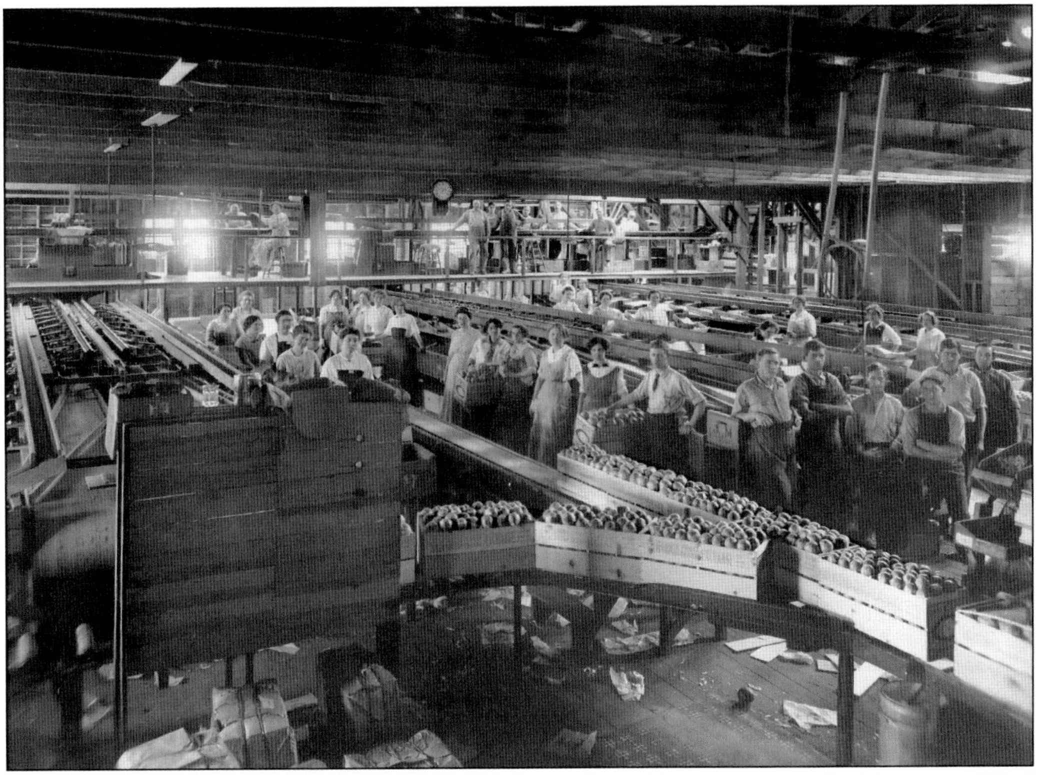

Inside the first Covina Citrus Association packing plant, the crew is posed with orange-filled crates carrying the famous Red label. It was printed by Los Angeles Western Lithograph Company. Standing center with his arm over a packing crate and wearing a shirt, tie, and suspenders is packinghouse manager Frank Anderson.

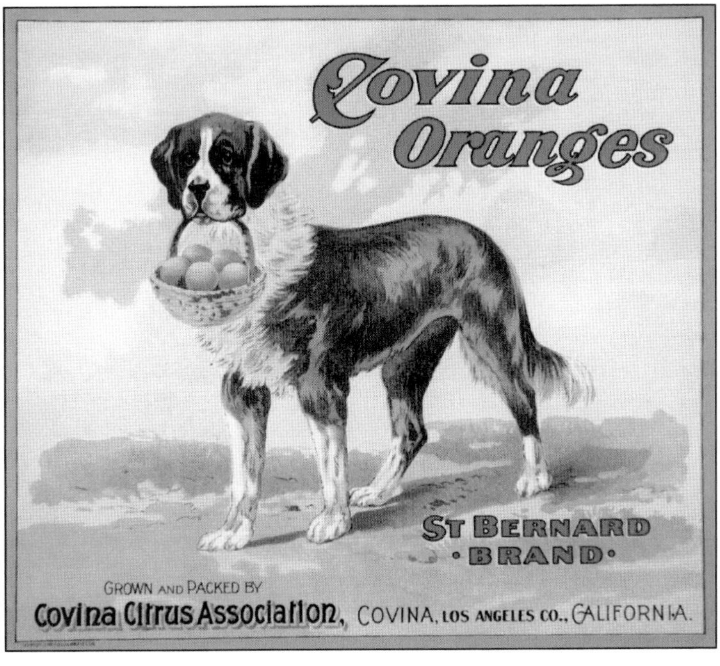

The St. Bernard label was designed and printed for the Covina Citrus Association by Los Angeles Schmidt Lithograph Company. It alludes to the Swiss Saint Bernard dogs that rescued stranded travelers in the Alps. Instead of carrying a keg of brandy around his neck, the dog is carrying a basket of oranges.

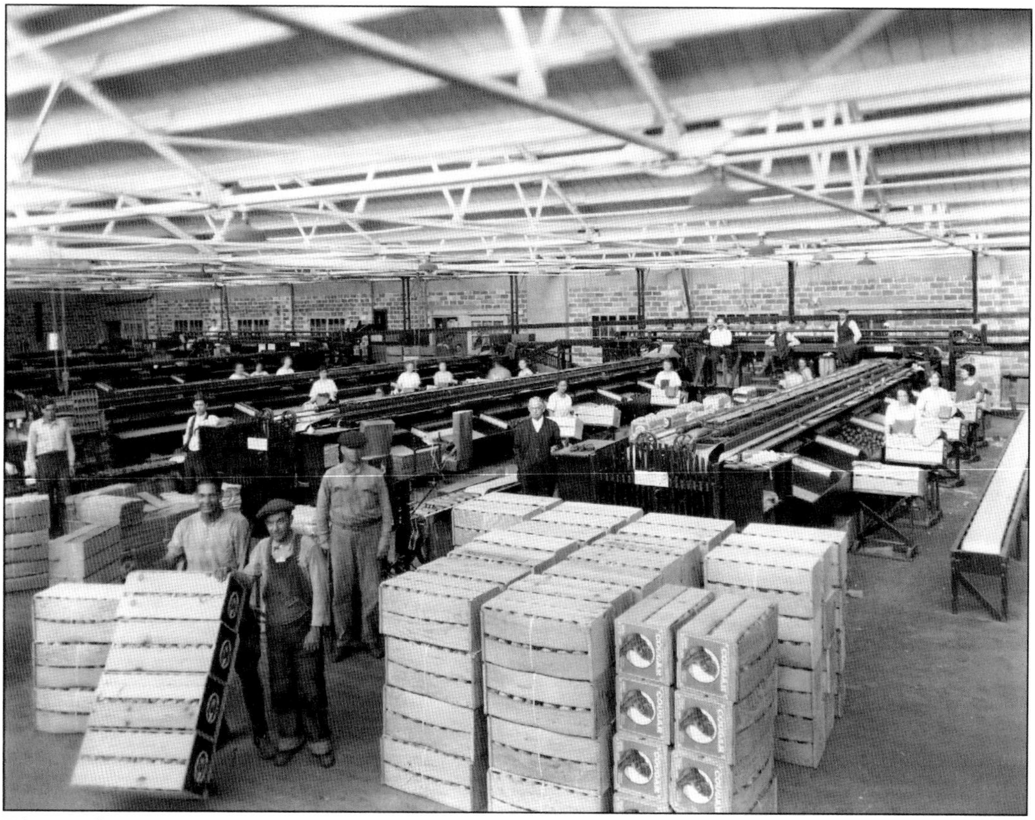

This is the interior of the new Covina Citrus Association packinghouse. The Cougar-labeled oranges are ready to go to the loading platform. The manager is standing in the center, wearing a dark suit with a vest watch chain and fob.

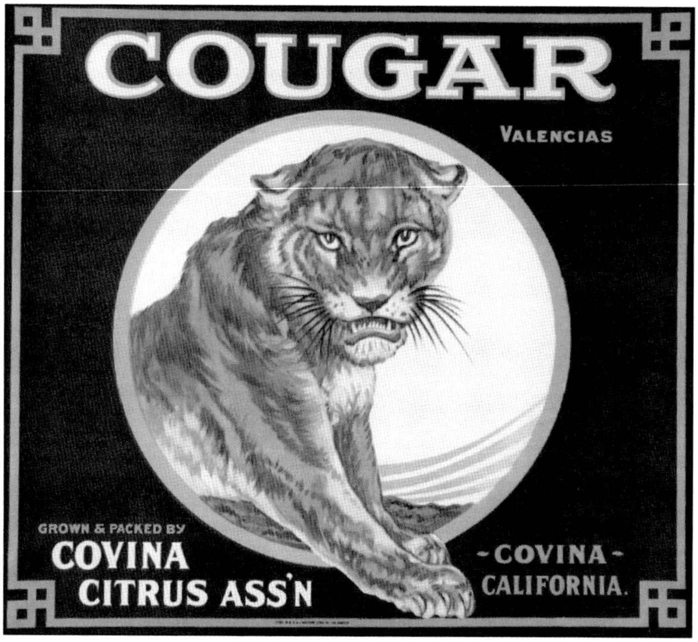

The Cougar Valencia label was designed and printed by Los Angeles Western Lithograph Company for the Covina Citrus Association. Notice the absence of the Sunkist trademark. All packinghouses that were members of the California Fruit Exchange shipped second-grade oranges using labels different from those used for Sunkist oranges.

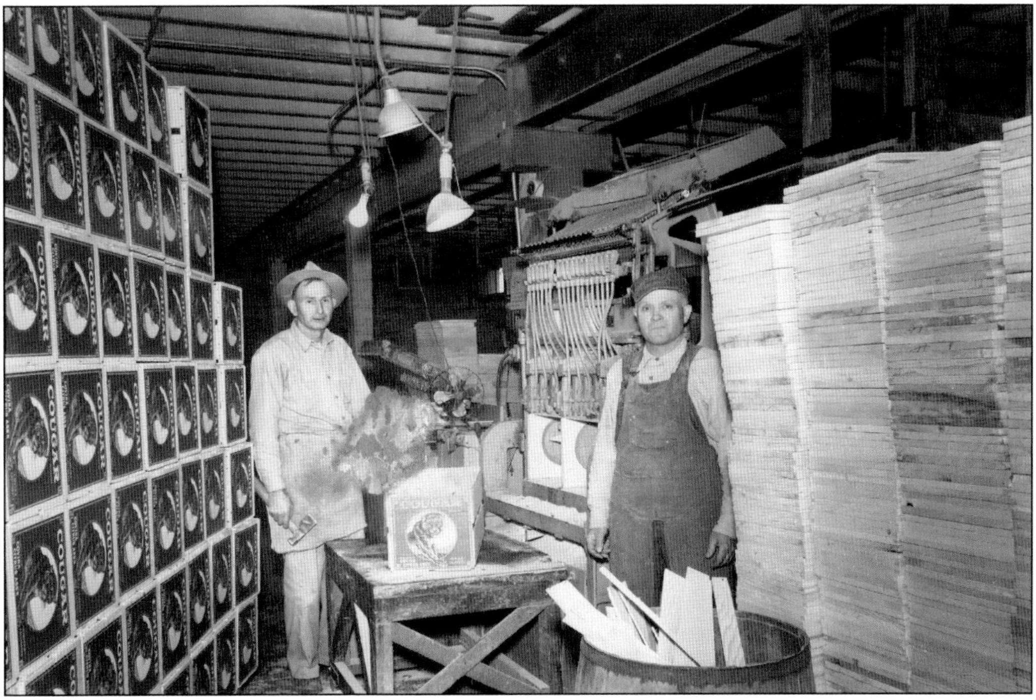

Covina Citrus Association box-makers Bill Wendt (left) and Bill Smith are standing in front of the box machine in the basement. Bill Smith worked for Covina Citrus from July 1927 to June 1952. The machine they operated could make 2,000 boxes a day. Notice the Cougar labels on the crates stacked to the left.

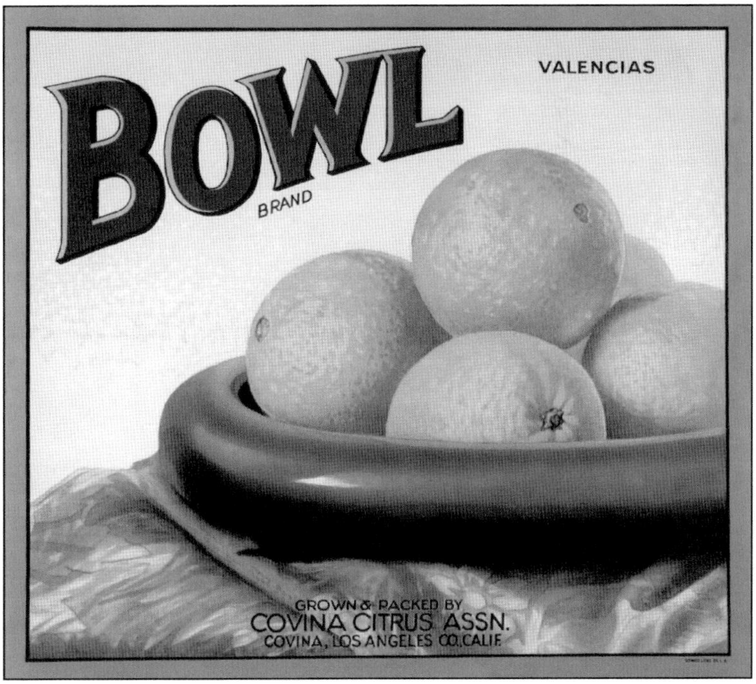

This Bowl Brand label for Valencia oranges was printed by Los Angeles Schmidt Lithograph Company for the Covina Citrus Association. The shipper is not mentioned on the label.

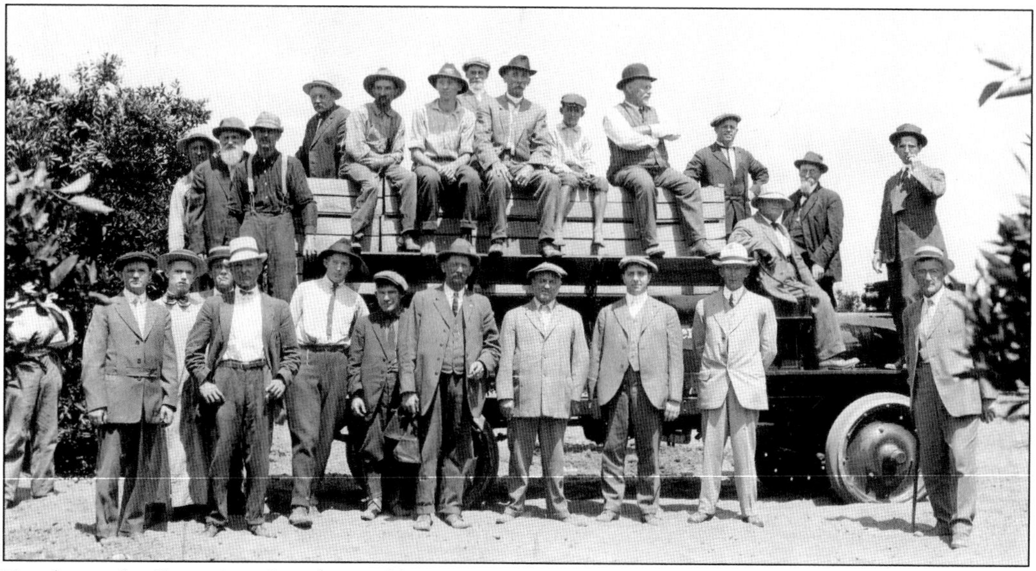

Trucks gradually replaced mules and horses. Covina Orange Growers Association members are posed with their new truck. In 1908, the *Glendora Gleaner* reported, "A motor truck with a speed of 12 miles an hour had been purchased by a Riverside grower for $4,000. The expense of operating this motor truck was considerably less than that of four teams."

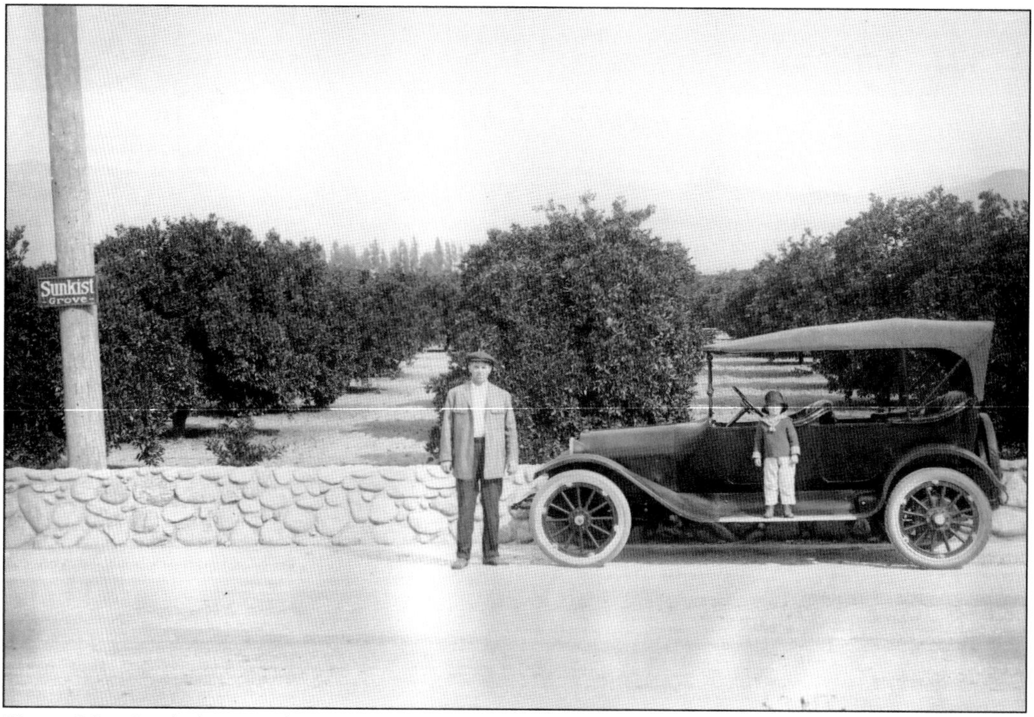

Henry Van Ginkel is standing next to his new car, and his daughter is standing on the running board. Proudly displayed behind him on the power pole is the Sunkist Grove sign. Growers with Sunkist-designated groves could charge higher prices for their citrus. After Sunkist began shipping citrus to Europe and Asia, their trademark became one of the most universally recognized in the world.

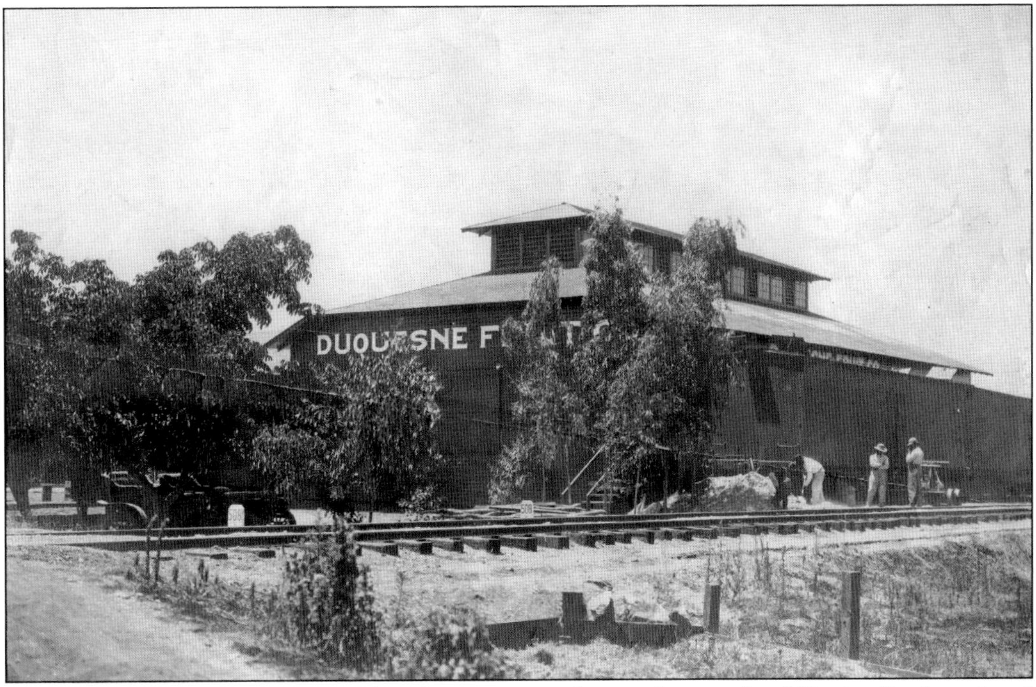

The Duquesne Fruit Company packinghouse was on the northwest side of Covina Boulevard and Sunflower Street in Charter Oak. It burned in 1927. In this photograph, a large, open touring car is shown next to a small car to the left. The company was named for Duquesne, a city near Pittsburg, Pennsylvania. Duquesne is located in the Golden Triangle, the area where the Monongahela and the Allegheny Rivers join to form the Ohio River.

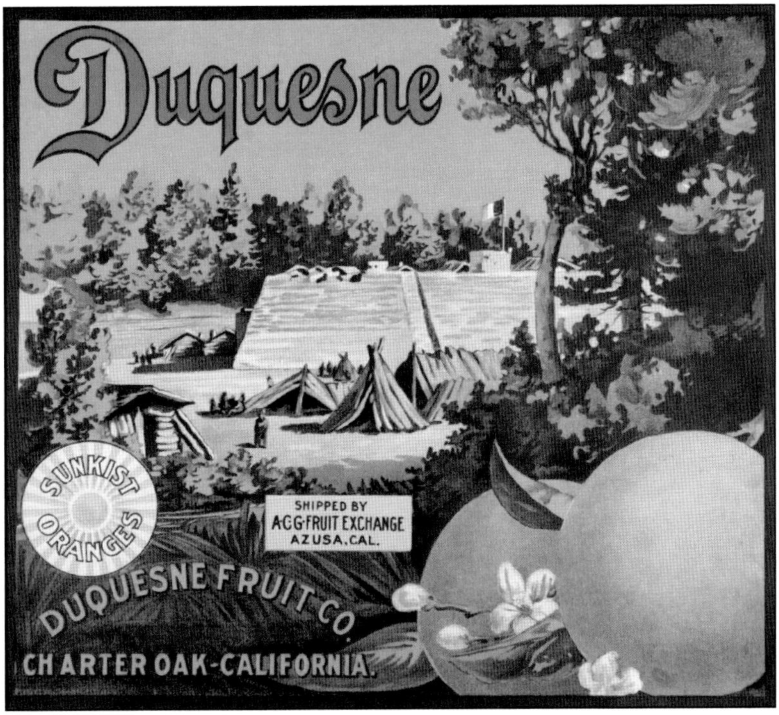

The Duquesne Fruit Company shipped under the Duquesne and the Old Mill Brand labels and used ACG Fruit Exchange Azusa as the shipping agent. The Sunkist medallion indicated the quality of the fruit and membership in the California Fruit Exchange. The small fort pictured on the label may be Fort Necessity where George Washington retreated in 1754–1755 after a battle with the French.

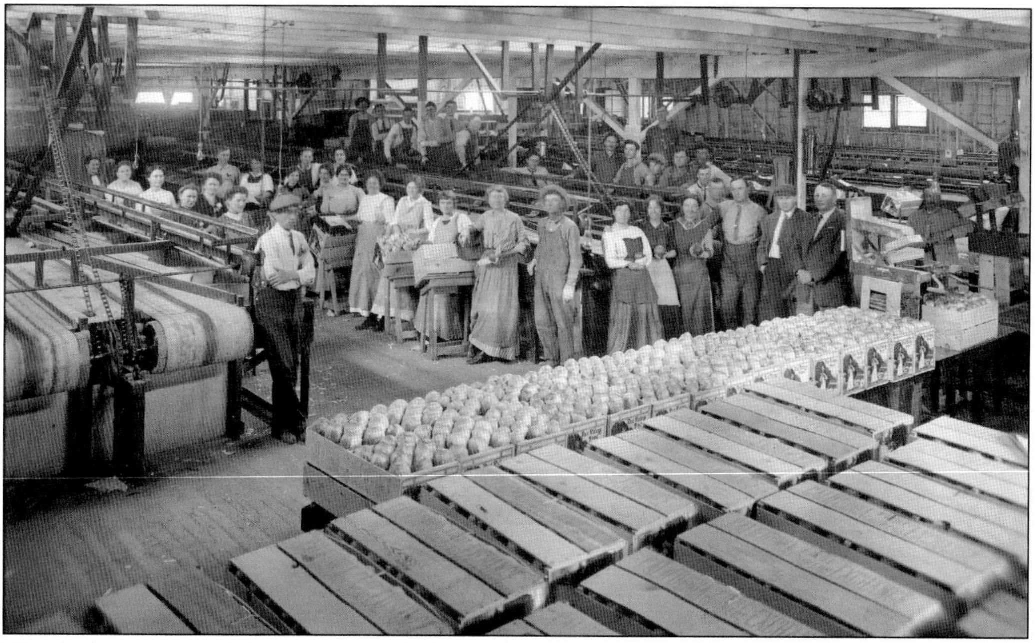

This is the interior of the Charter Oak Citrus Association packinghouse in 1925. The two men wearing jackets are bookkeeper Randal Chew (left) and manager William Hoogandyke. Notice the Little Red Riding Hood labels on the packing crates.

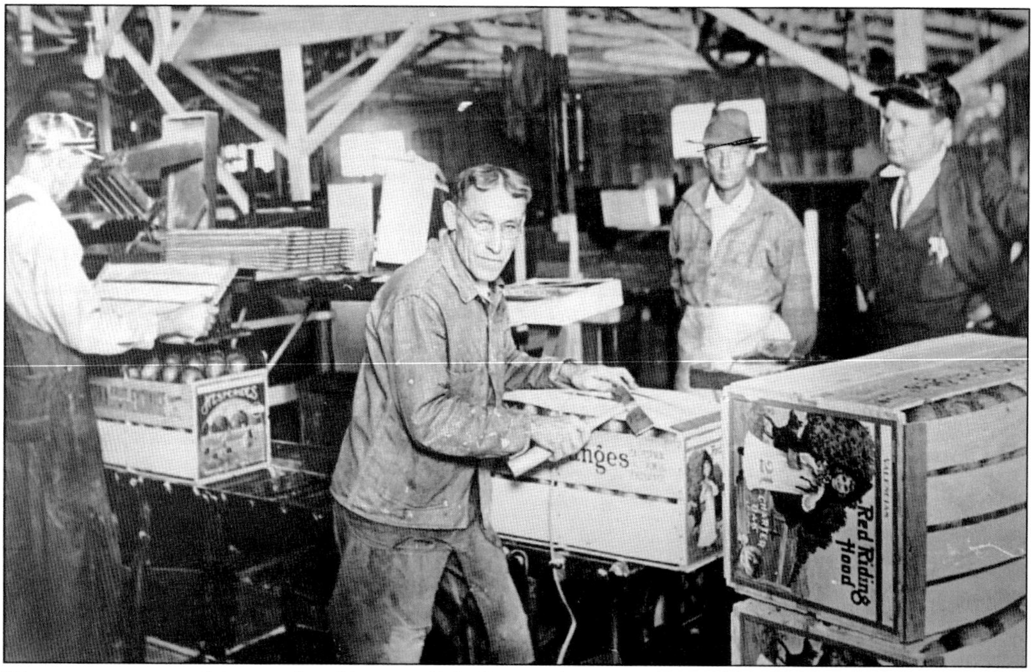

These men are working in the Charter Oak Citrus Association packinghouse readying boxes of Hesperides- and Red Riding Hood–labeled oranges for shipping. The machine in the background made the boxes. From left to right are foreman Carl Smith, stripper Otto Frederick (with hatchet), car loader Glen Morgan, and pressman Fred Johnson. The Charter Oak Citrus Association formed in 1902.

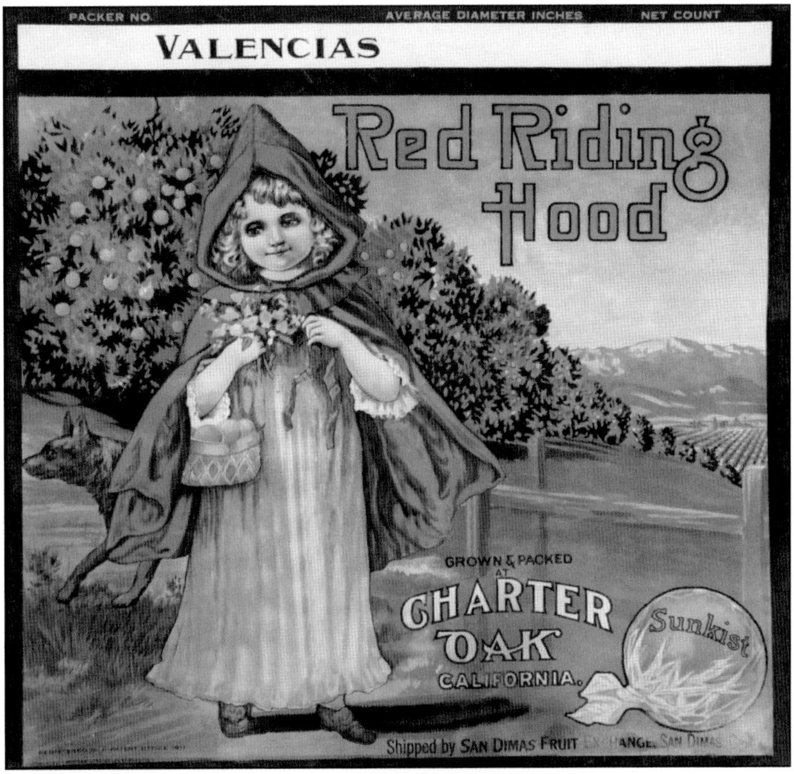

The Red Riding Hood Brand label was printed by Los Angeles Western Lithograph Company for the Charter Oak Packing Company. It has the wrapped-orange Sunkist trademark, and the shipping agent was the San Dimas Fruit Exchange. Red Riding Hood was part of a label series based on nursery rhymes. The familiar character is shown in a citrus grove with a basket of oranges to take to grandma.

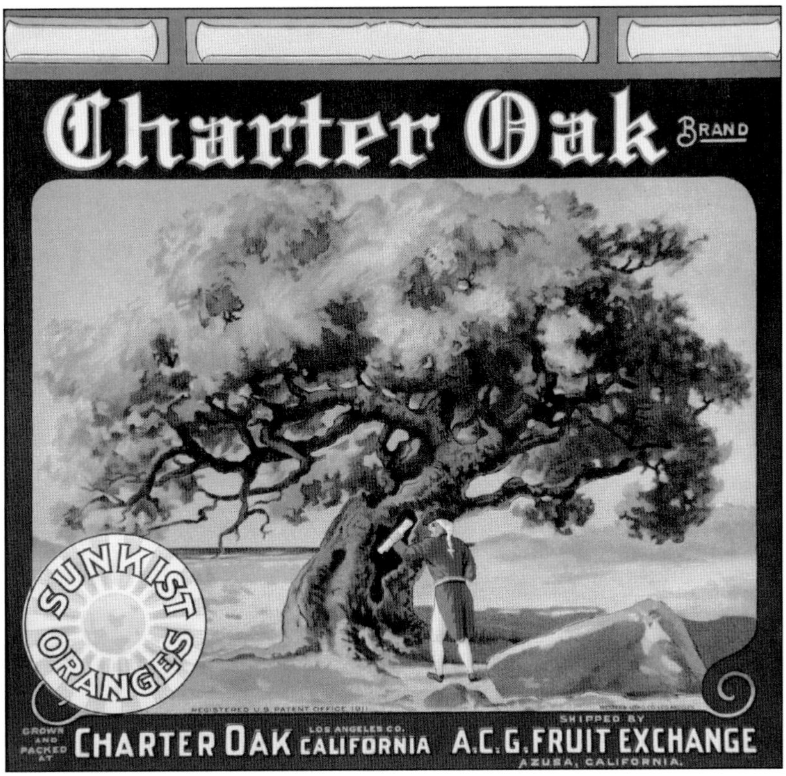

The Charter Oak Brand label was printed by Los Angeles Western Lithograph Company for the Charter Oak packinghouse. The illustration refers to the famous Charter Oak in Connecticut where a Colonial soldier hid the Connecticut Charter during the American Revolution. San Dimas Fruit Exchange was the shipping agent.

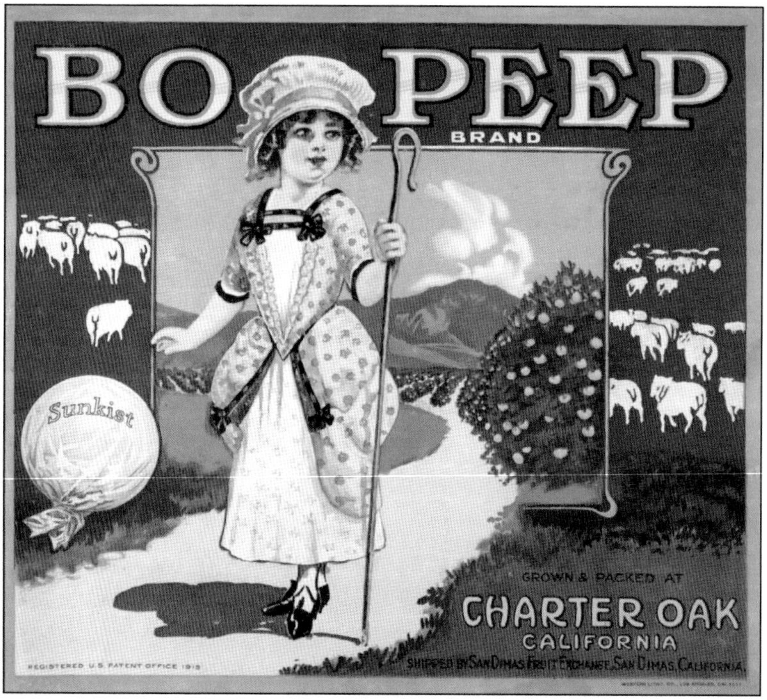

Part of the nursery rhyme series, the Bo Peep label was designed and printed by Los Angeles Western Lithograph Company for the Charter Oak packinghouse. Bo Peep, shown with her shepherd's crook, has lost her sheep in a citrus grove. San Dimas Fruit Exchange was the shipping agent.

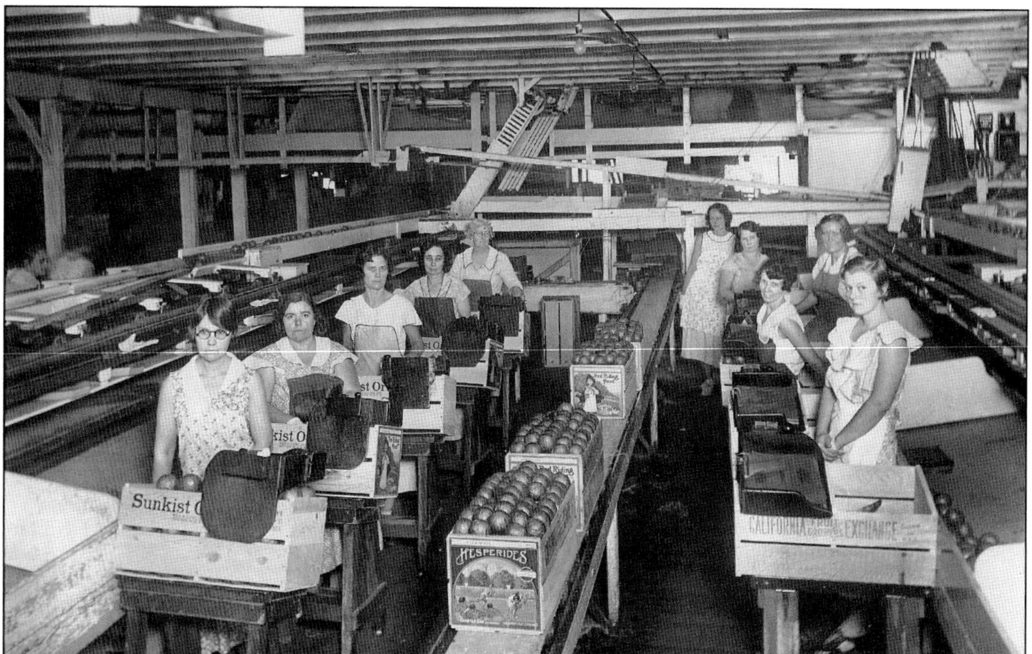

In this photograph of the Charter Oak Citrus Association packinghouse from 1932, Red Riding Hood and Hesperides labels appear on the packed orange crates. The left packing line includes, from front to back, Sally Morgan, Millie Martin, Ella Bonter, Mrs. Laws, and unidentified. The right packing line includes, from front to back, Mrs. Marsh, Hazel Coffee, and three unidentified women.

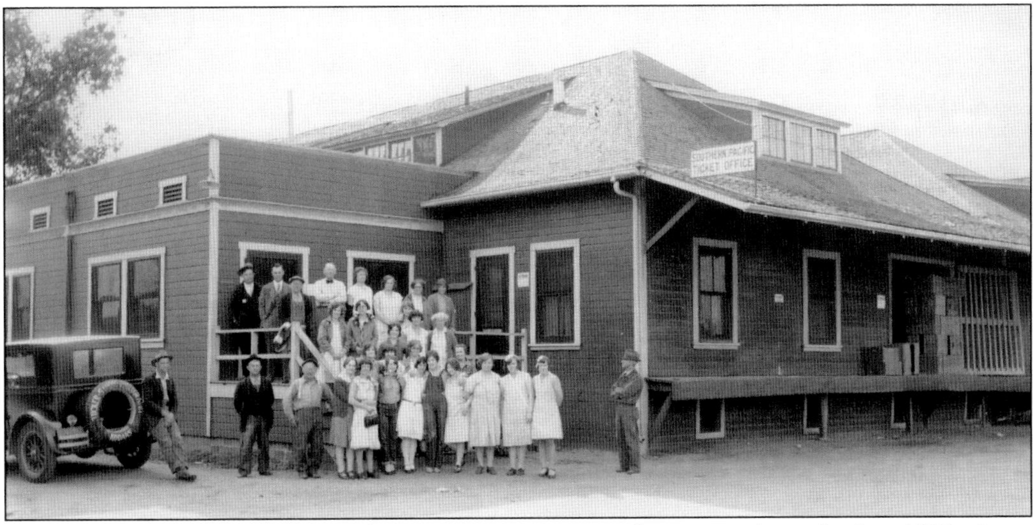

A photograph of the Charter Oak Citrus Association packing crew from the late 1930s shows how women's clothes and hairstyles have changed. Notice the two women wearing blue jeans and overalls in the first row (third and fifth from left). Their noticeably different attire remains a mystery. Silent film star Dale Fuller, who owned Fuller Ranch on Sunflower Avenue, was the only other woman in Charter Oak to wear overalls or jeans at the time.

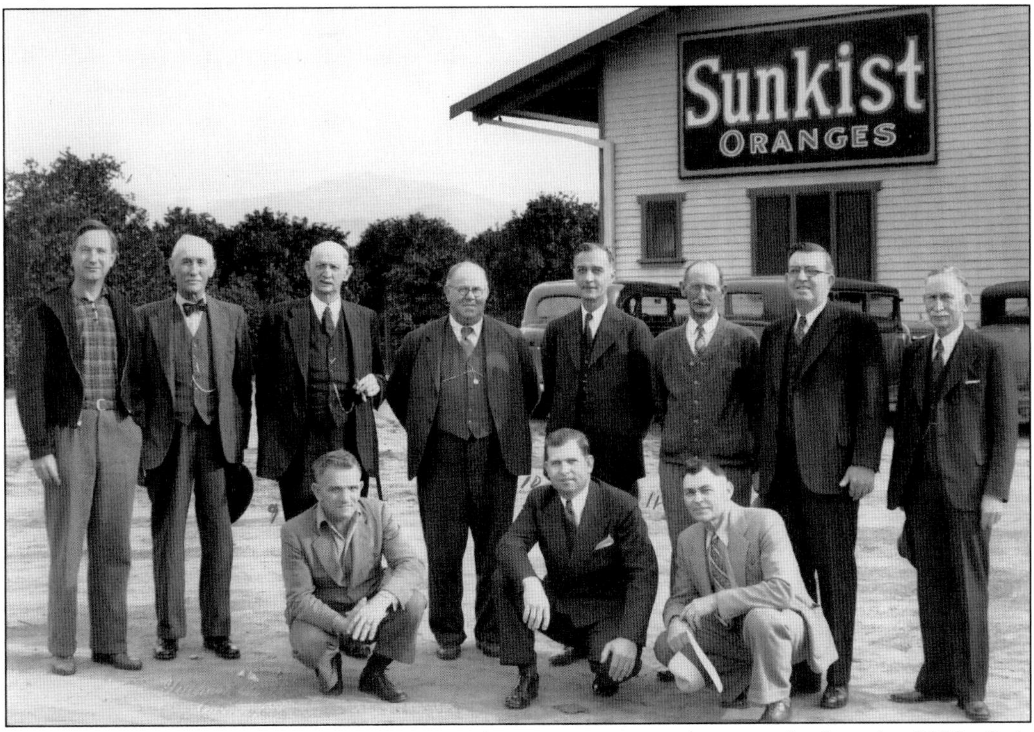

Directors of the Covina Orange Growers Association were photographed in the 1920s. It is interesting that the Housers, once independent packers, are now directors of a Sunkist distributor. Pictured are, from left to right, (kneeling) John Kock, Henry Hepner (manager), and J.D. Pitzer; (standing) Ralph White, John Houser, Harv Houser, Ed Blanchard, Max Leonhardt, J.W. Houser, C.W. Hunter (bookkeeper), and William Griswold.

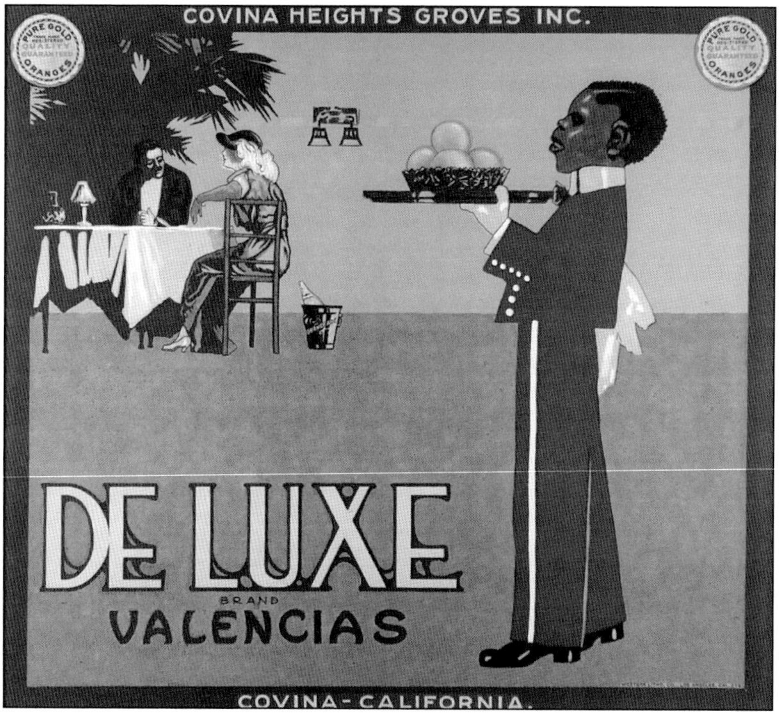

The Deluxe Brand Valencia label was designed and printed by Los Angeles Western Lithograph Company for independent packer Covina Heights Groves, Inc. The label carries the Pure Gold orange seal. A waiter is serving oranges to an elegantly dressed couple.

Los Angeles Western Lithograph Company also printed the Deluxe Brand label for El-Bar-Dor Citrus, Inc., Valencias. In this label, the same waiter from the Covina Heights label above is serving a different couple. The man and the woman have been updated and are wearing more modern clothes. The label has the tissue-wrapped Sunkist trademark and advertises that the fruit has been Brogdexed to reduce decay and retard shrinkage. The reappearance of the waiter is an excellent example of how printers kept generic label designs—or in this case, partial designs—in stock.

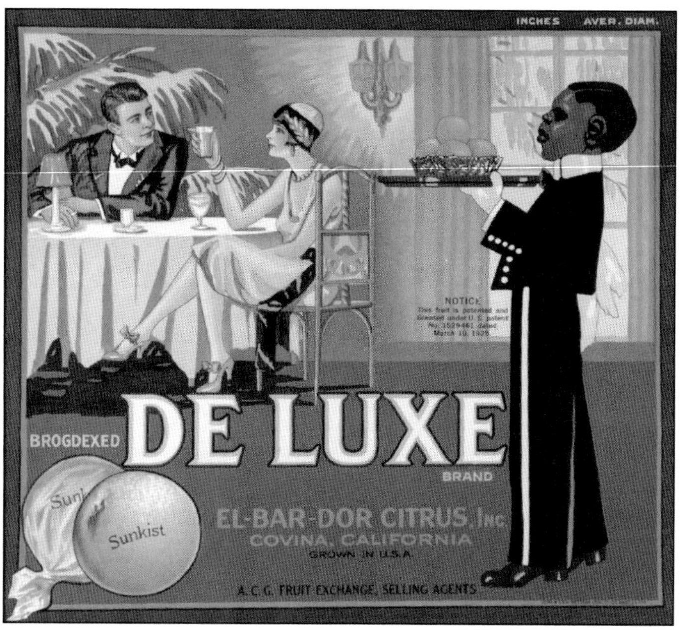

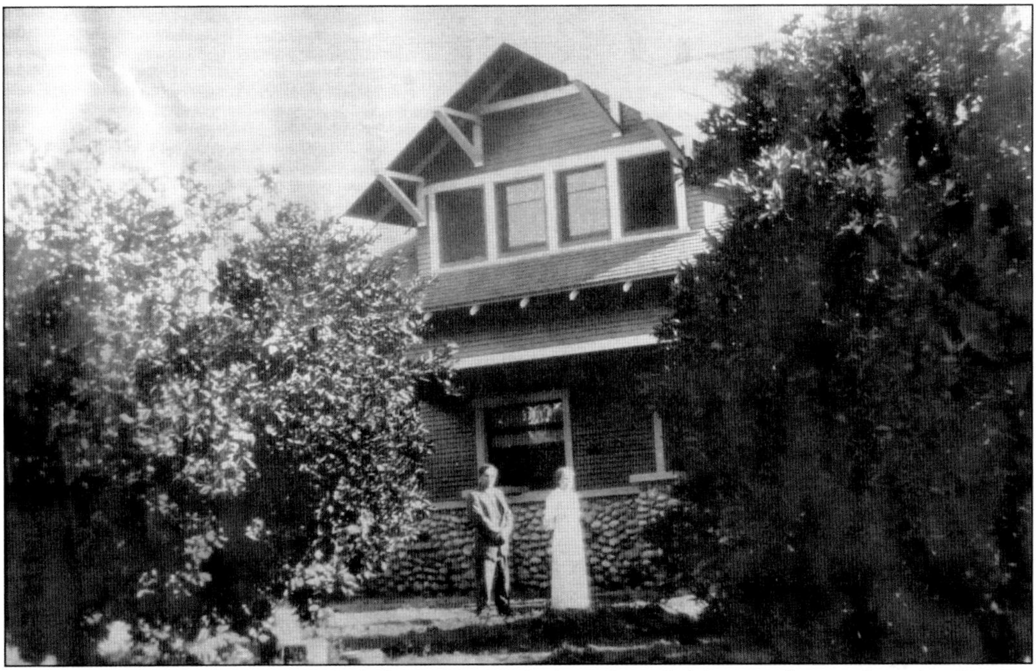

Catherine and Leonard Daniels left Kansas with their five children and moved to the Covina area. Their youngest son, Leonard Daniels Jr., was born after they arrived. The Daniels family established 60 acres of citrus east of Hollenbeck Avenue on land that is now part of West Covina. Their home, shown, is the present-day site of Sacred Heart Church, Covina. They were Covina Citrus Association stockholders and active in the Baptist church.

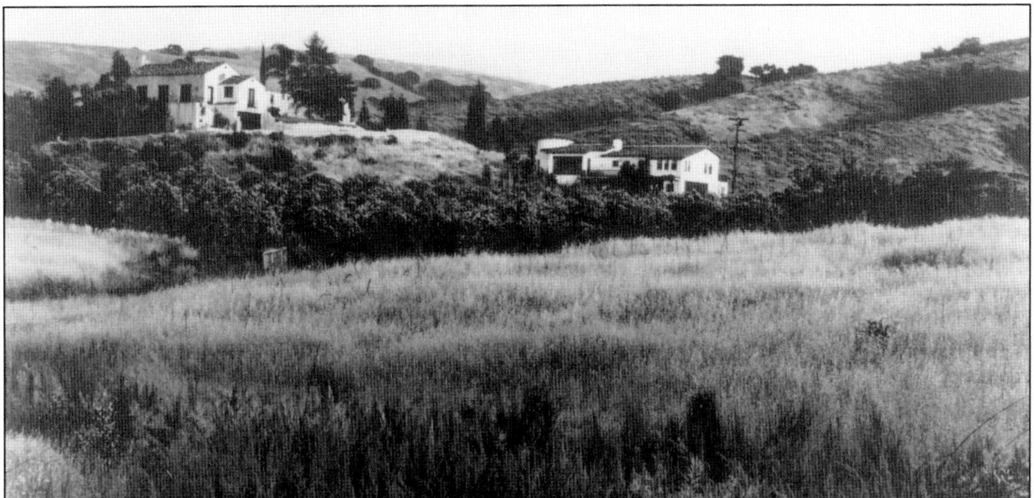

In 1918, Judson Reeves purchased part of the Hollenbeck Ranch and named it the Covina Highlands. He planted 290 acres of oranges, which were destroyed in the freeze of 1922. By 1924, three homes had been built in the highlands along Cameron Avenue by Anna and Ben Thorpe, Lark Ellen, and the Benedict family. The Thorpes' Spanish-style home in the photograph was called El Guadalajara. They collected California plein air paintings. Lark Ellen, who was the sister of Anna Thorpe, built a 350-seat echo bowl below her home for concerts, operas, and Easter services.

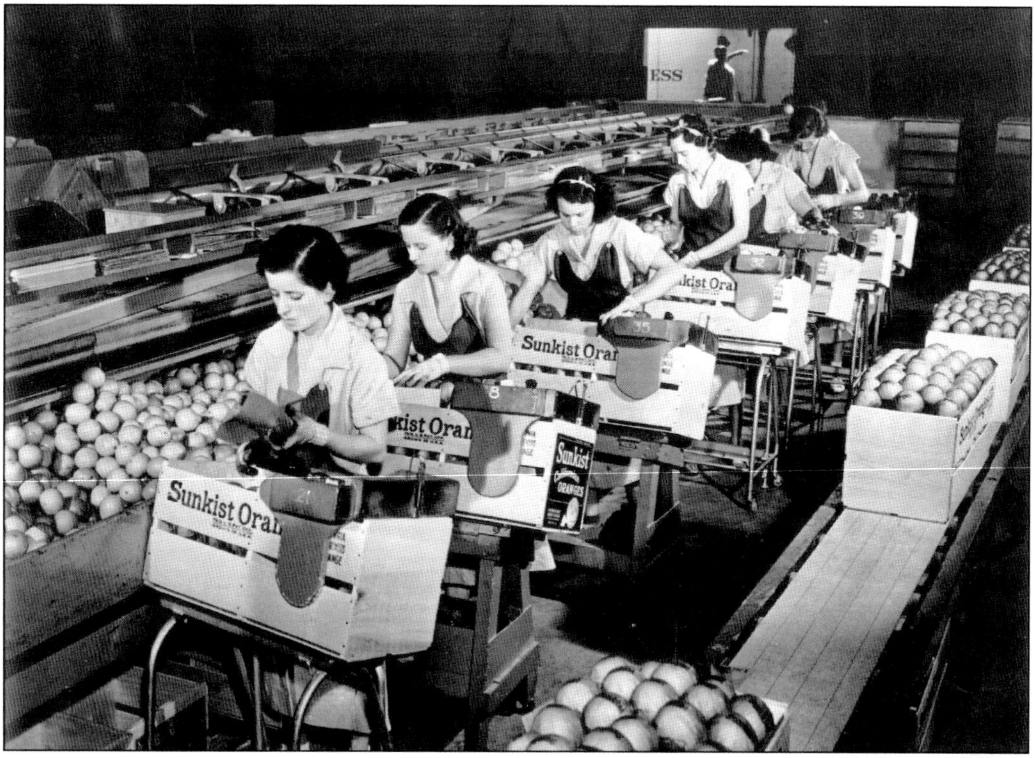

In 1935, the women wearing Damerel-Allison uniforms are packing Sunkist-labeled oranges. They are wearing gloves to keep from bruising the fruit. Each orange was wrapped in tissue paper. The best fruit marketed through the California Fruit Exchange was stamped with the Sunkist trademark, which is known worldwide. However, Covina fruit of equal quality was shipped under labels such as Blue Goose and Pure Gold.

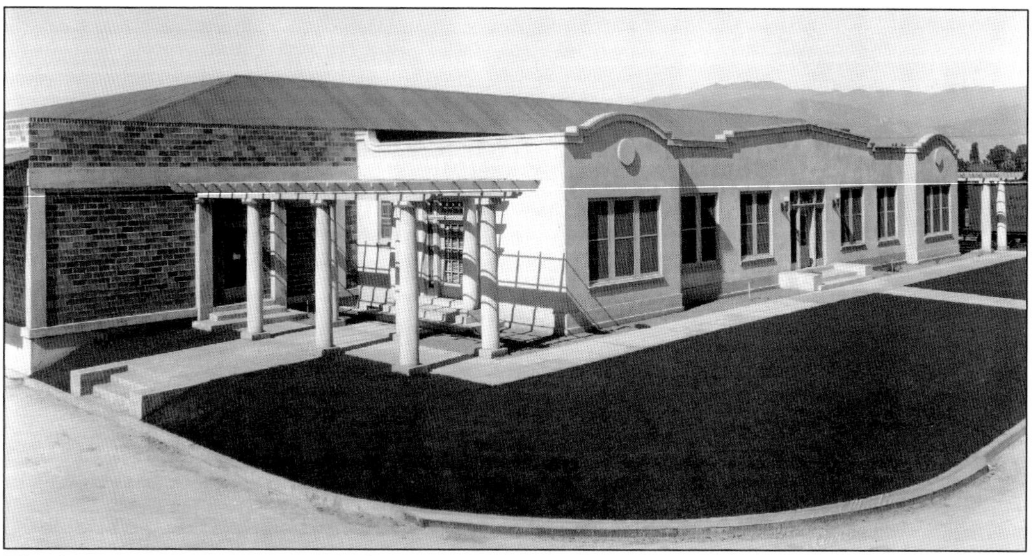

The Covina Citrus Association built a new packinghouse, pictured, in 1920 to replace the original, which was built in 1895. It was located on Third Street near the Southern Pacific Railroad tracks. The brick packinghouse had modern offices.

Seven

New Citrus Products

After the booming citrus years of the 1920s, the stock market crash in 1929 was a great shock to valley citrus ranchers. Many lost their life savings. During the Great Depression, citrus prices held up at first, but later dropped to or below the cost of production. In 1931, the Glendora bank closed. In Covina, J.C. Penney, Broadwells Department Store, and the Palace of Sweets closed. Growers had problems getting the short-term loans from banks needed to finance production. In 1933, the Farm Credit Act established production-credit associations to help American farmers with production costs. The federal government also tried to regulate citrus shipments at a time when supply was plentiful but demand low. The Independent Packers Mutual Orange Growers and the American Fruit Growers objected to these rules because they felt their quotas were unfair. The *Covina Argus* reported that some valley ranchers were letting their fruit drop because they could not afford to pick it and send it to the packinghouse.

The county established the Serra Sewing Center where unemployed valley women were paid to make clothes distributed through the Los Angeles County Welfare Department. Azusa was not as badly affected by the depression. The Rainbow Angling Club was successfully launched in 1930, and the construction of Morris Dam provided jobs when employment in the area was at an all-time low.

After 1940, the citrus market began to pick up; in 1944–1945, revenues hit an all-time high. The new citrus by-products business had become an $11-million-a-year industry. Many valley residents give credit to Harry Damerel of Covina for keeping the citrus industry alive with the development of citrus by-products during the late 1930s and early 1940s. Although frozen orange juice was first developed in Florida, the Damerel-Allison Company became the major California producer. Truckloads of culls from California and Arizona arrived daily at their juice plants. They also pioneered the delivery of fresh orange juice by the Arden Diary in Los Angeles County.

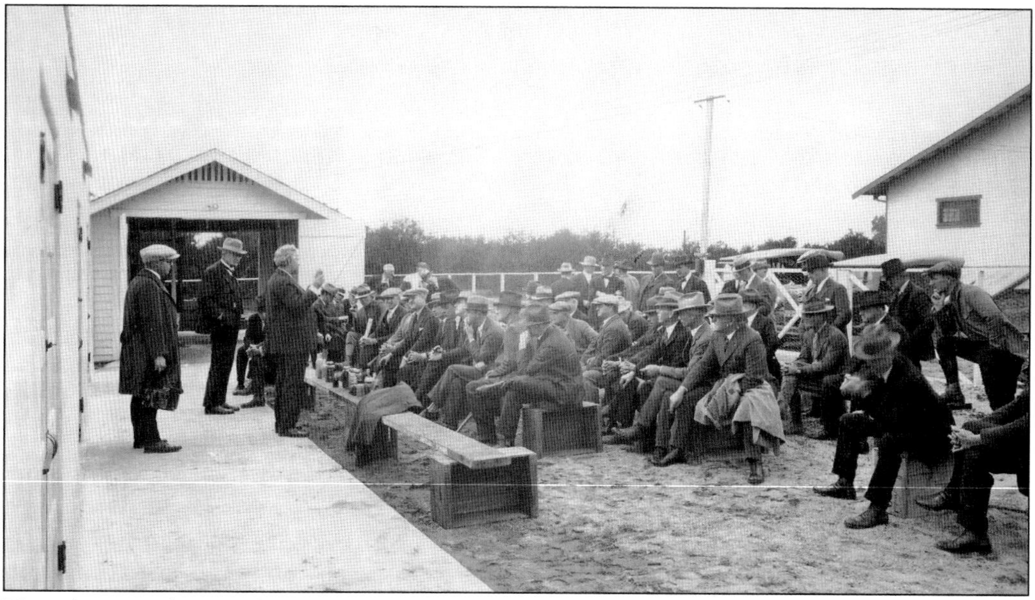

This photograph was taken in 1917. The chemist who ran the Covina Citrus Association juice laboratory is giving a lecture to members of the Farm Bureau on the use of chemicals in citrus production. He is illustrating his lecture with specimen bottles and jars arranged on the first bench. Because his laboratory was too small, the lecture was given outside. The men are sitting on improvised benches made from wooden planks and field crates. The man behind the chemist on the left is carrying a camera.

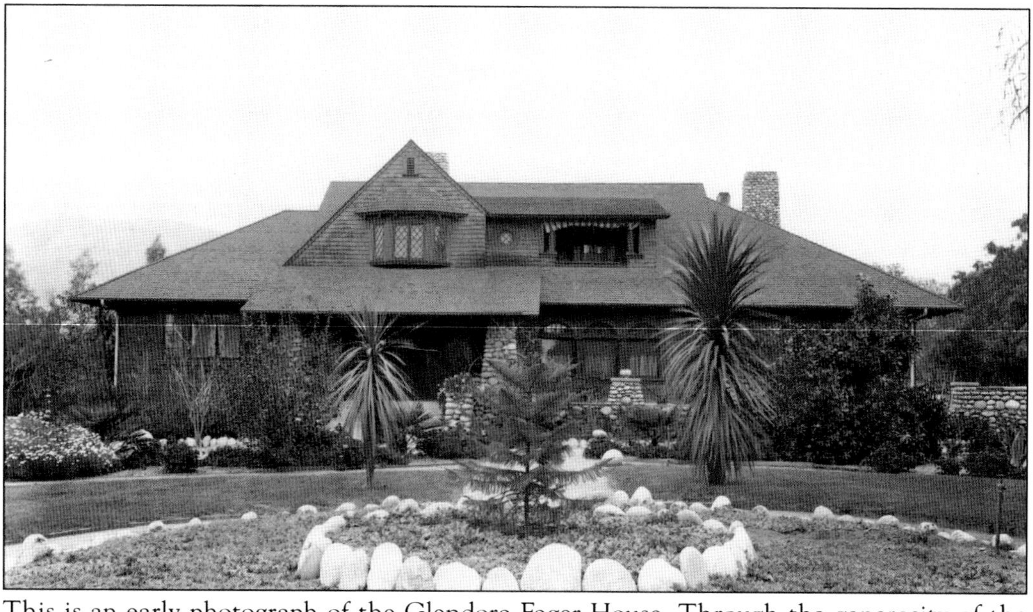

This is an early photograph of the Glendora Fager House. Through the generosity of the Ludiovich and the La Fetra families, it was saved from demolition and became the Glendora Senior Center. In 1935, Mary Elizabeth and Clem La Fetra founded the Rain Bird Corporation. They used the first impact sprinkler, invented by their neighbor Orton Englehart, to build their company into the leading manufacturer of irrigation systems in the world. (Courtesy of Glendora Historical Society.)

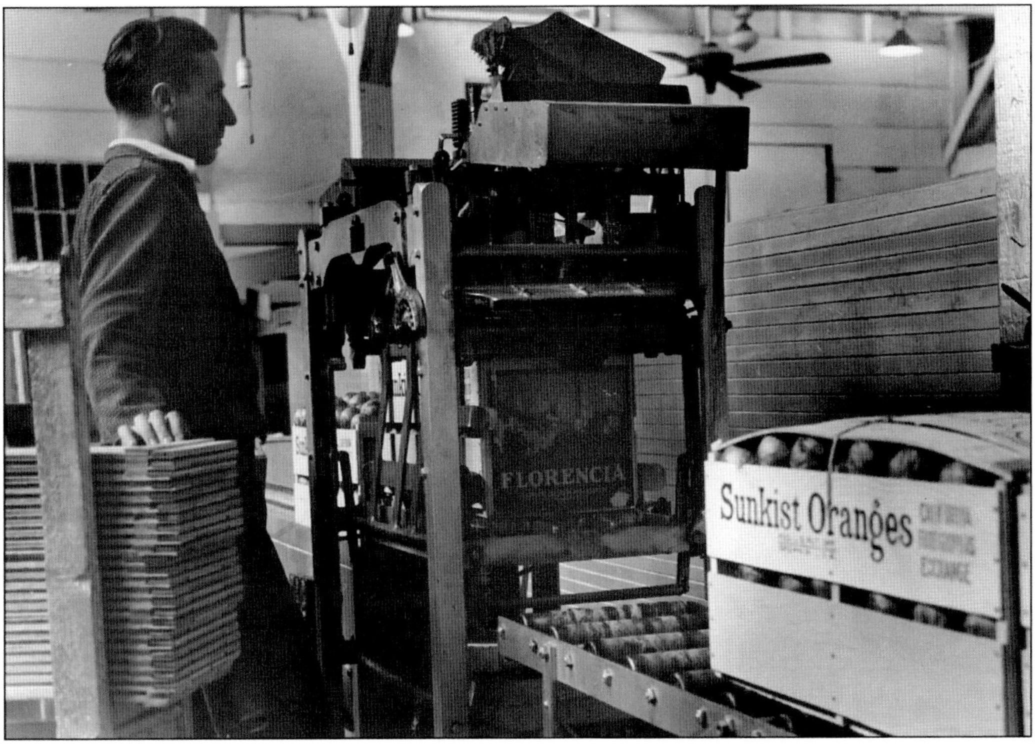

This is the crate-cover operation at the Damerel-Allison citrus packing plant on Azusa Avenue. In one operation, this machine puts on the crate cover without injuring the fruit, puts on a steel strap over the cover, stamps the size and number of oranges in the box, and records the number of boxes that the machine handles. Before the lid goes on, the fruit in the boxes sits about two and a half inches above the box.

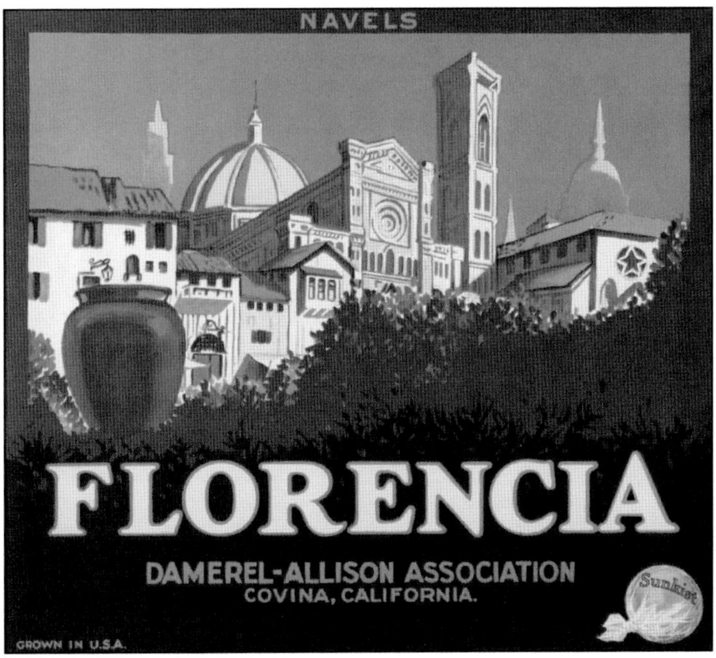

The Florencia Brand label was printed for the Damerel-Allison Association. They also shipped under the Athena label. The photograph on page 113 shows a box with the Florencia Brand label attached. The label has the tissue-wrapped Sunkist trademark and shows a fanciful view of Florence, Italy.

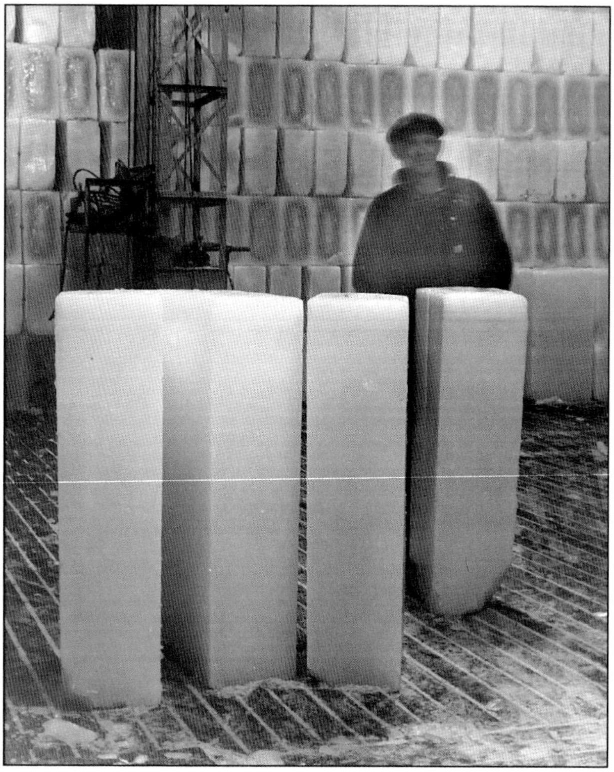

At the Damerel-Allison plant in Covina, four 320-pound cakes of ice are on their way to the stacker in the ice storage room. The 360-ton-capacity room was insulated with four inches of cork on the sides and bottom and five inches overhead. The temperature was kept at 26 degrees Fahrenheit, and a machine stacked the ice and moved it down.

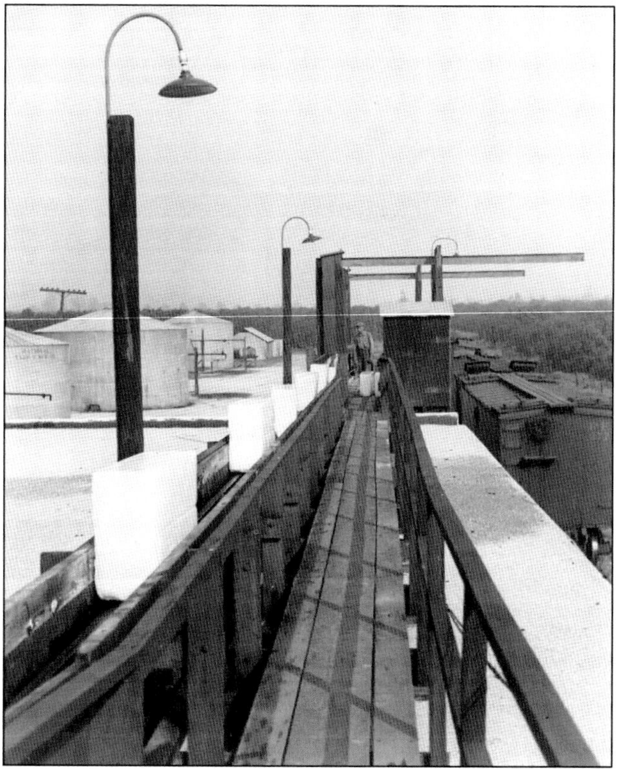

Ice is on its way to refrigerator car bunkers via a conveyor belt. Thirty minutes were required to load seven tons of ice into each car. Behind the conveyor belt are oil storage tanks with a 100,000-gallon capacity. Beneath each tank was a cement underground storage tank of the same capacity. The oil stored was low-grade, 28-percent-gravity oil; it cost 3¢ a gallon and was used for smudging citrus.

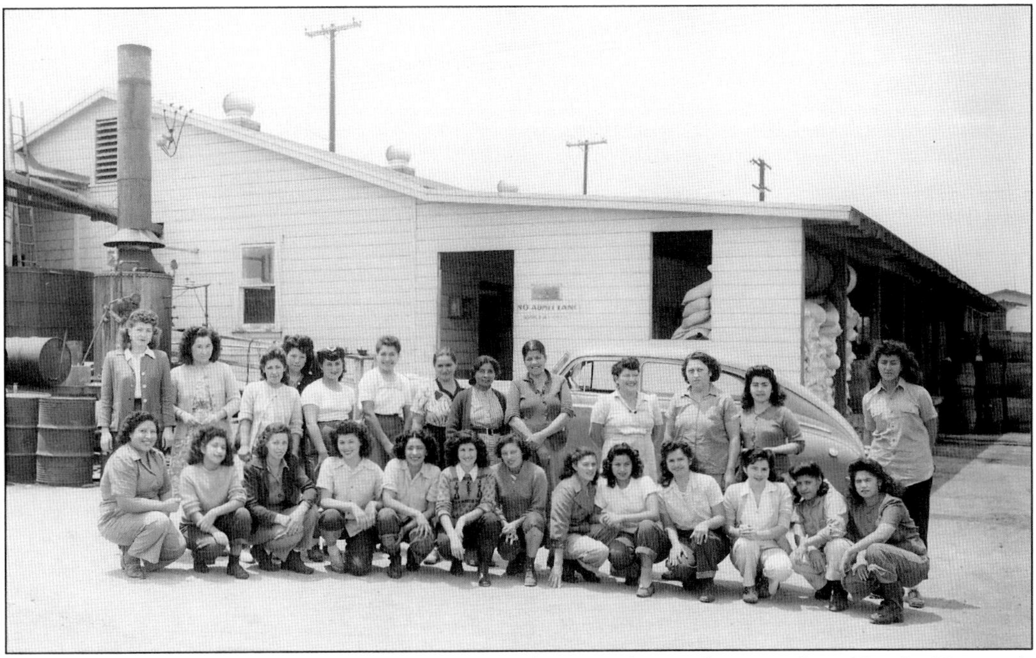

Pictured is the female processing crew for one of the citrus by-products produced by the Damerel-Allison Company. The warehouse was connected to Plant No. 2. The orange peels from juice extraction were processed using various machines and preserved in brine. Notice that the younger women are wearing practical pants for work instead of skirts.

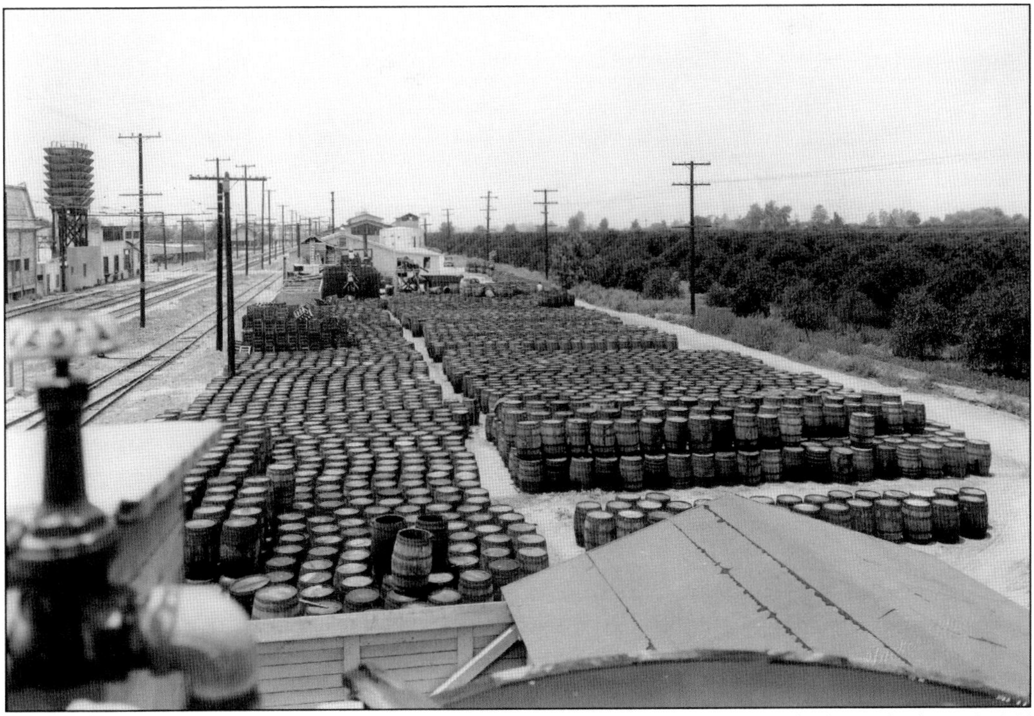

Outside the Damerel-Allison Company warehouse was a brine-curing yard where the barrels full of processed orange peel waited until they were shipped.

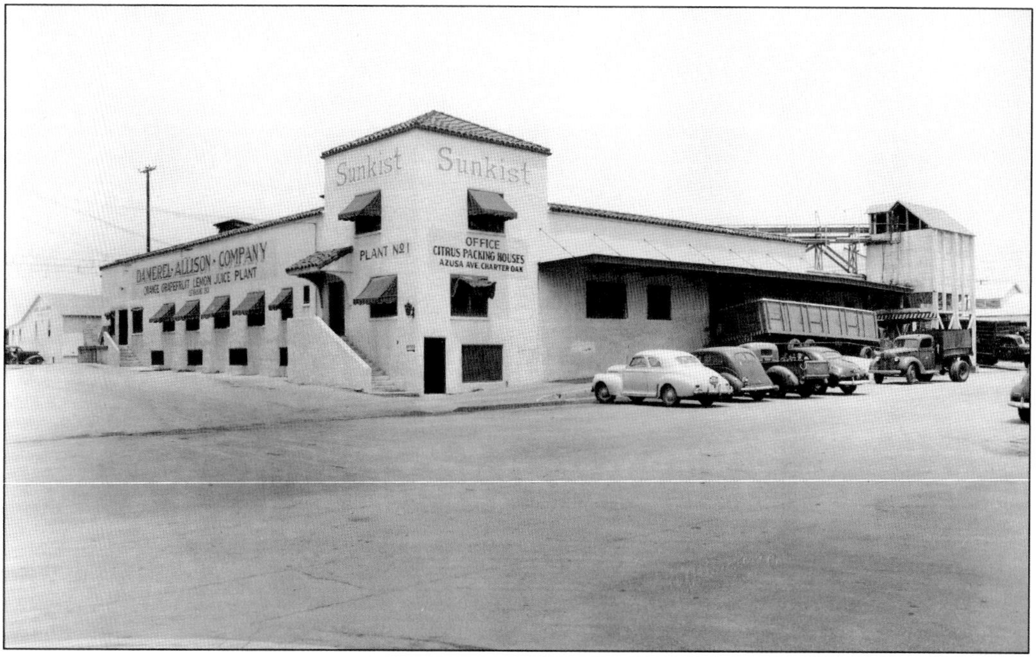

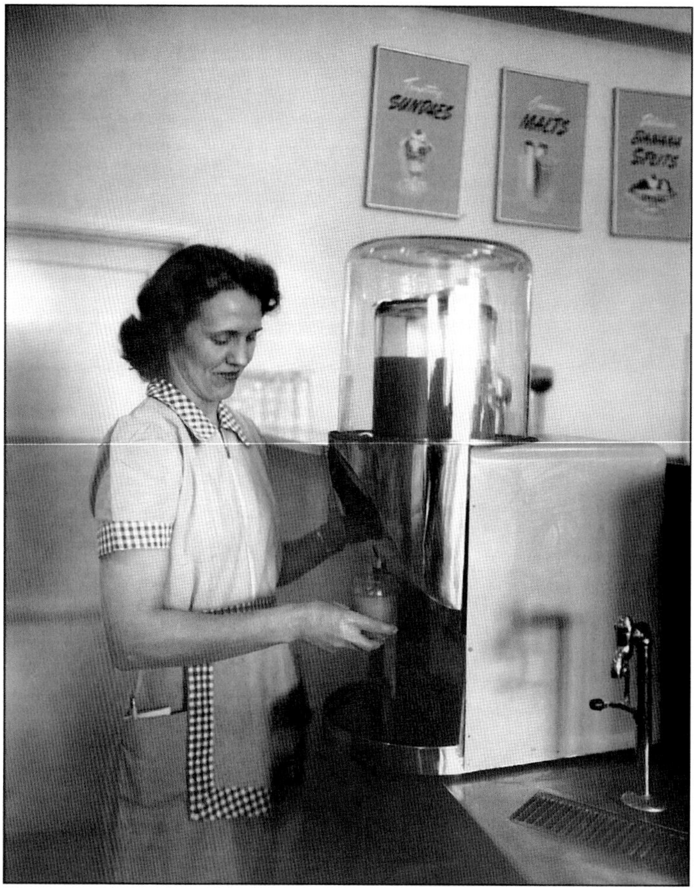

Harry Damerel and Herman Allison, citrus growers and businessmen, joined to create the Damerel-Allison Company, which sold citrus juices and by-products. This photograph shows Plant No. 1 on the northeast corner of Front and First Streets in Covina where they processed orange, grapefruit, and lemon juice.

The Damerel-Allison Company not only sold juice but also juice dispensers. A waitress is shown here serving a glass of orange juice from a dispenser in Pop Jenks' Sugar Bowl, a teenage hangout on Citrus Avenue that was made famous by the syndicated cartoon strip *Harold Teen*. The Sugar Bowl served hamburgers, malts, and banana splits, and now they had fresh orange juice.

This style of D&A juice dispenser was made for movie theaters. The first one was installed in the lobby of the Covina Theater.

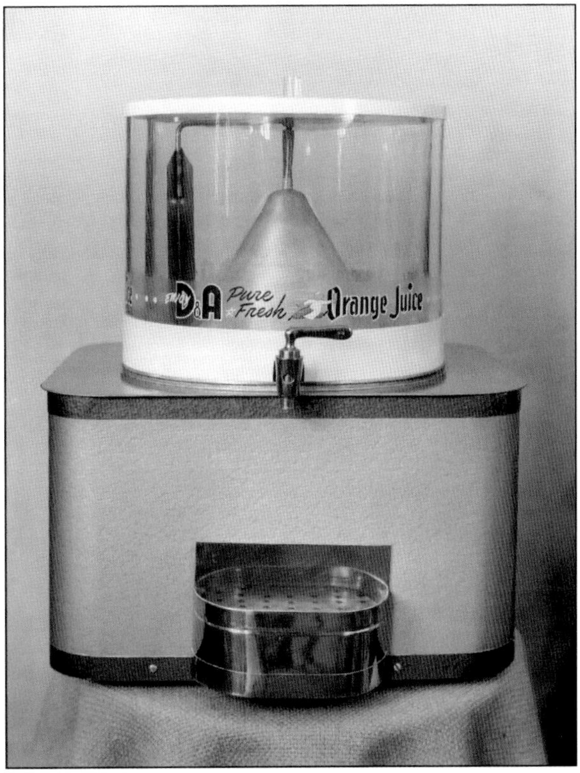

This photograph shows a line of refrigerated juice dispensers in one of the Damerel-Allison plants.

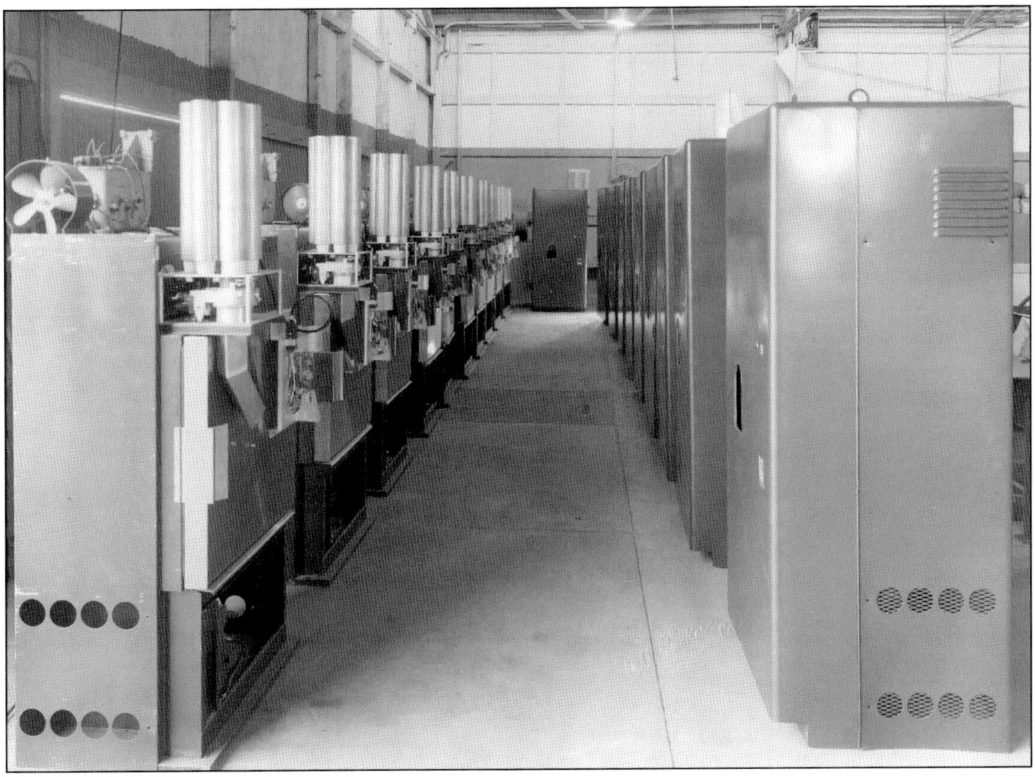

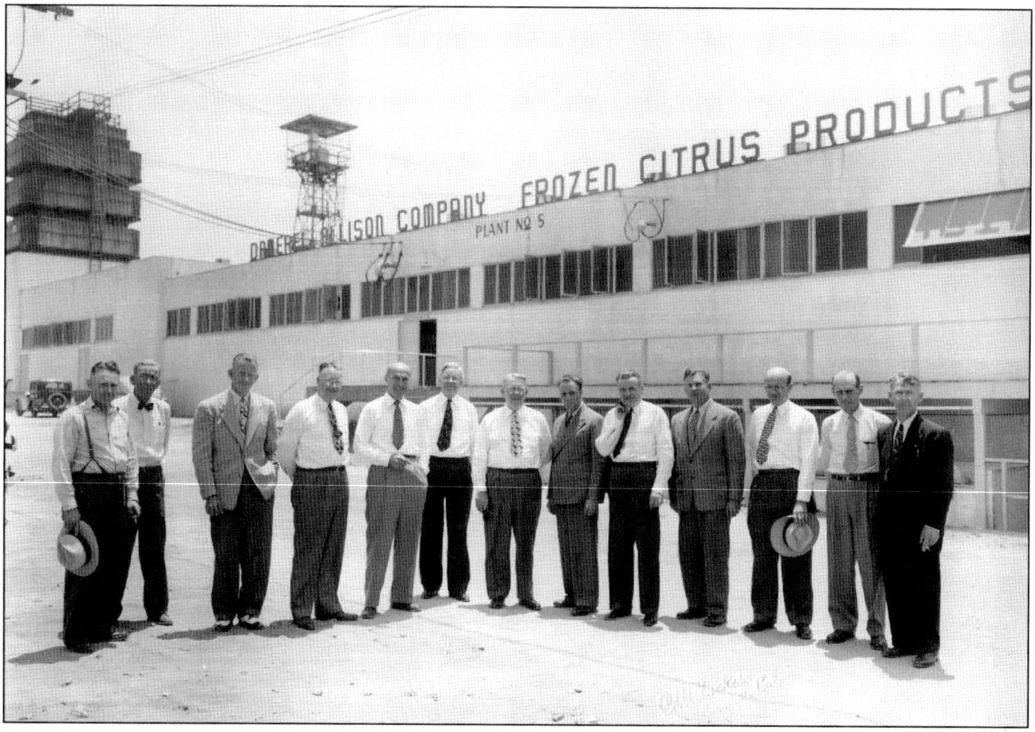

The plant managers of the Damerel-Allison Company are photographed in front of the greatly enlarged Plant No. 5. This plant was converted and enlarged in 1946–1947 to handle the processing of frozen juice. It became a tremendous operation, drawing citrus culls from all over California and Arizona. Hundreds of people were employed at the plant. James Hodges is fourth from left.

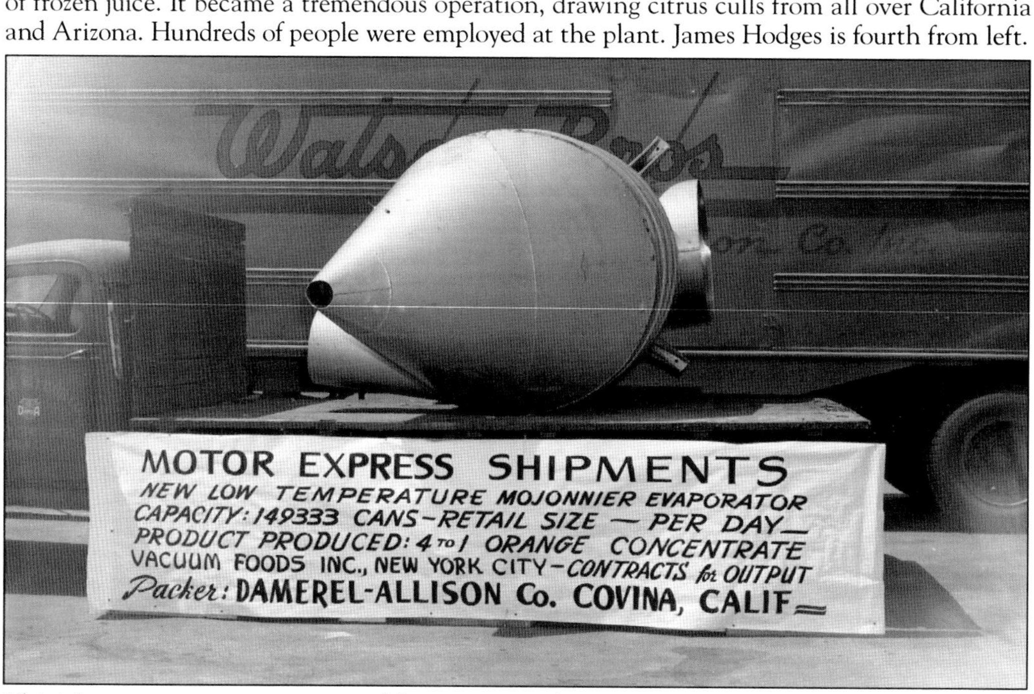

This Mojonnier evaporator was used for processing frozen juice. The Damerel-Allison Company always had the latest equipment.

Eight

Citrus Era Ends

In the 1940s, the Covina Valley citrus industry began its gradual, but not smooth, decline toward the end. Industry survival depended completely upon a constant supply of fruit. The packing, shipping, and marketing done by Sunkist and the independent packers began to shut down. Packinghouses began to close, and fruit exchanges merged.

In 1943, Thomas Merton Griswold was recognized for his years of service to the California Fruit Exchange (Sunkist) as a director and manager of the Covina Fruit Exchange since 1898. Elbert Black Griffith of Azusa and W.W. Wilson of Glendora were recognized as having served as directors since 1925. Mrs. Charles Warren of Glendora received a special award for having been an exchange member since 1883. The 500 new acres of Valencias planted in the Covina Highlands were producing. Damerel-Allison built a new frozen juice plant that employed several hundred people in Covina. They also started producing orange-meal feed for diary cattle and citrus-peel products.

In 1949, when viewed from the San Jose Hills to the San Gabriel Mountains, the entire valley was still a vast citrus grove with three small communities set among the trees. There were no freeways or shopping malls. The last stoplight on Garvey Boulevard was in El Monte; from El Monte to Kellogg Hill, there were no other stops. Policeman hired by West Covina to chase speeders caught numerous motorists driving above the speed limit on this straight-shot stretch of road.

In 1945, valley citrus had started to die from a new disease. In 1951, Dr. Robert C. Dickson discovered the quick decline that was destroying trees in Brazil and California was caused by a virus transmitted by melon aphids or infected rootstock. The disease, known as the citrus tristeza virus (CTV), occurred when the phloem tissues died in the bud union area, and photosynthesis could not take place. Orange trees grown on sour rootstock (*Citrus aurantium*) were especially susceptible to the disease. As groves died, residential subdivisions grew. Agricultural land was rezoned and taxed at residential property rates, making citrus ranching financially unfeasible.

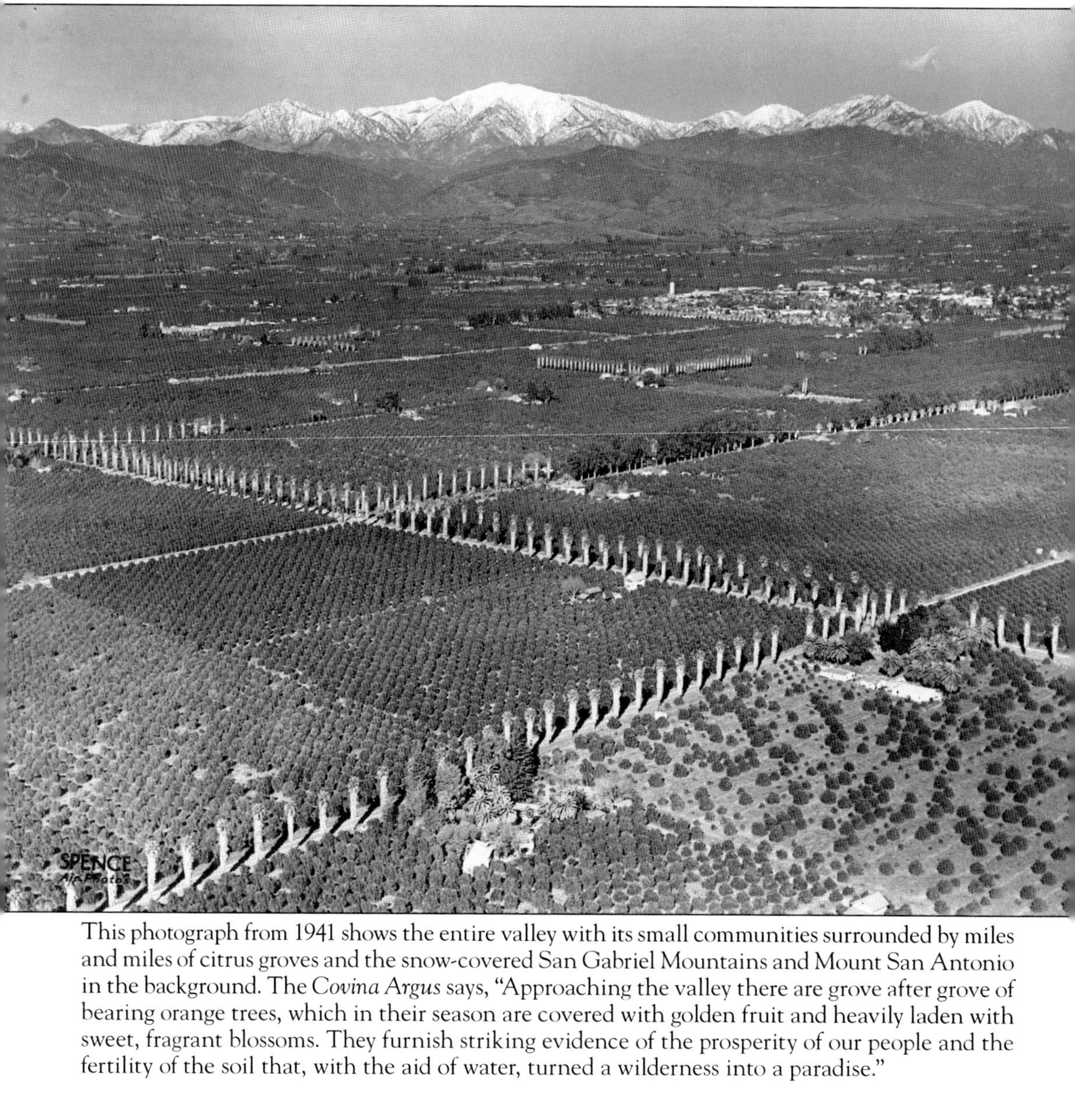

This photograph from 1941 shows the entire valley with its small communities surrounded by miles and miles of citrus groves and the snow-covered San Gabriel Mountains and Mount San Antonio in the background. The *Covina Argus* says, "Approaching the valley there are grove after grove of bearing orange trees, which in their season are covered with golden fruit and heavily laden with sweet, fragrant blossoms. They furnish striking evidence of the prosperity of our people and the fertility of the soil that, with the aid of water, turned a wilderness into a paradise."

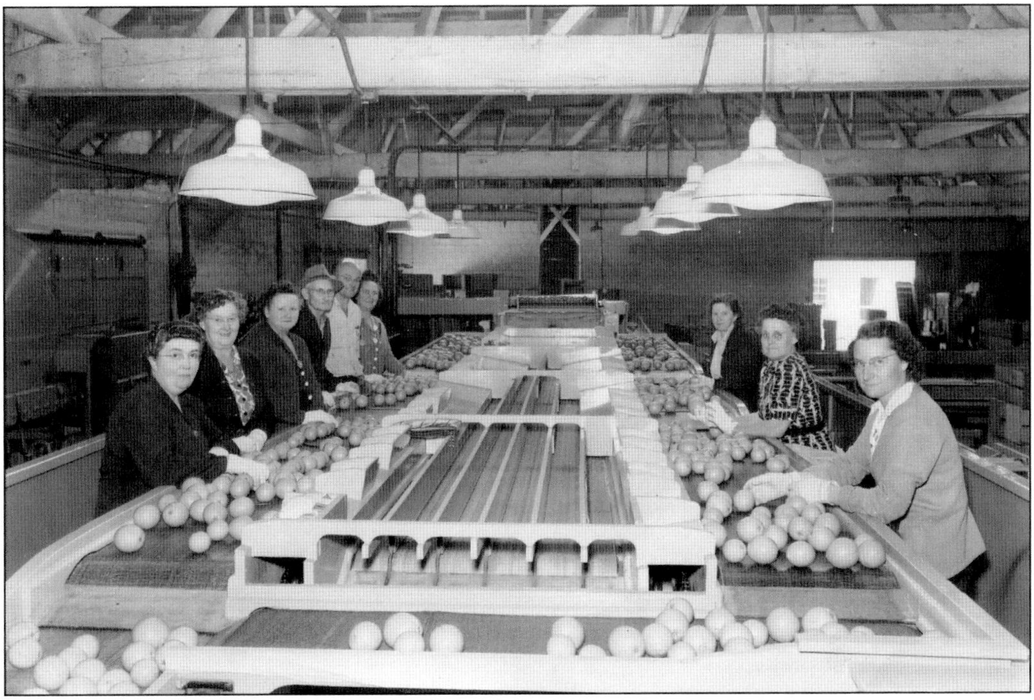

The men and women in this photograph had worked for the Glendora Citrus Association packinghouse. They met for a reunion. Notice the very modern packing equipment. Even as the valley citrus era was ending, packinghouses continued to install the latest equipment. They even installed x-ray machines to check for frost damage. (Courtesy of Glendora Historical Society.)

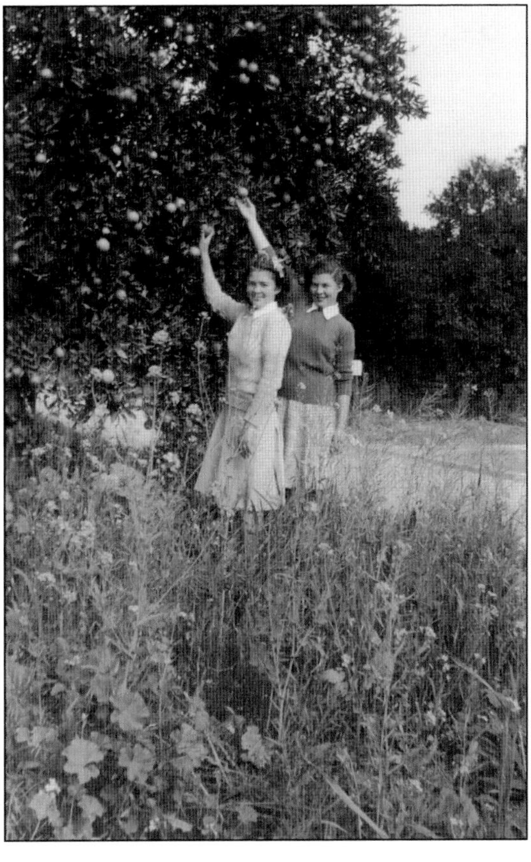

Sisters Marilyn (left) and Arline Dial are picking oranges in a grove managed by their father, Damon Dial. Henry Hepner, manager of the Covina Orange Growers Association packinghouse, owned the 10-acre grove on Garvey Boulevard between Lark Ellen and Azusa Avenues. This grove had not been cultivated because it was part of an experiment to see if non-cultivated groves would survive CTV. This grove did.

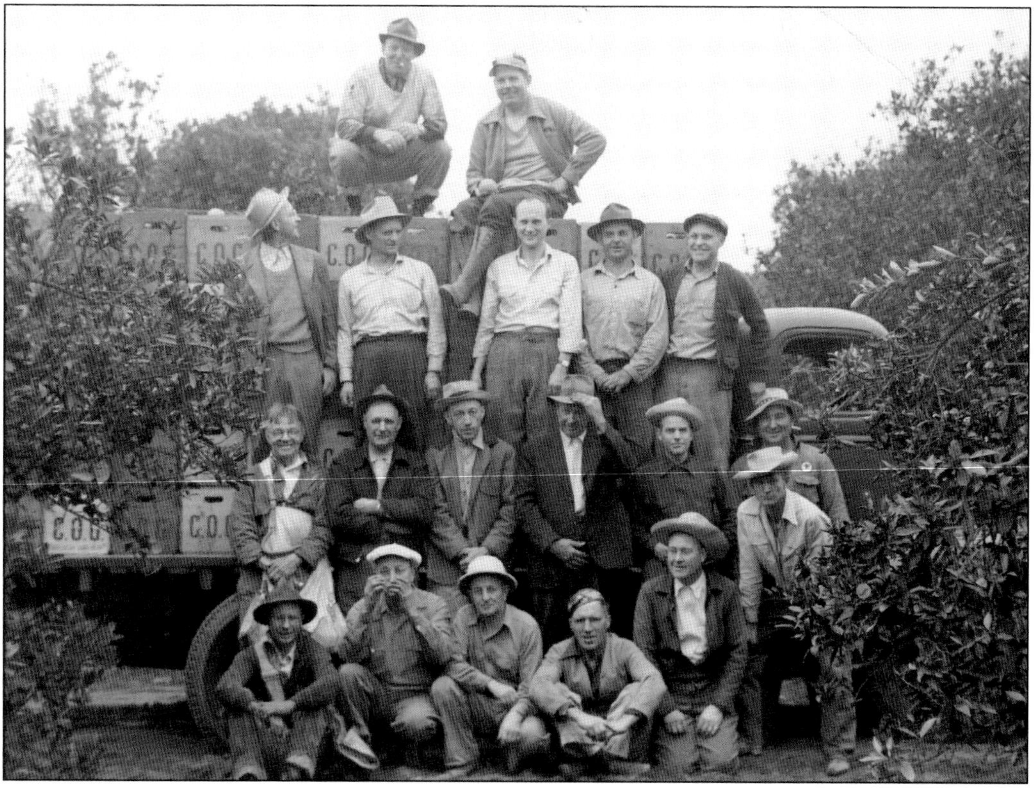

In 1944, during World War 11, Covina Lions Club members picked oranges to raise money. Pictured are, from left to right, (first row) Earl Clover, Tom Finch, Guy Lang, John Tallman, James Hodges, and Lowell Miller; (second row) Bert Owens, Al Ferrier, Ben Jenks, Bob Estep, Clyde Price, and George Ross; (third row) Oscar Yeager, Thomas Reed, Corwin Hoffland, Jay Pitzer, and Guy Harris; (fourth row) Scotty MacDonald and Bob Grondahl.

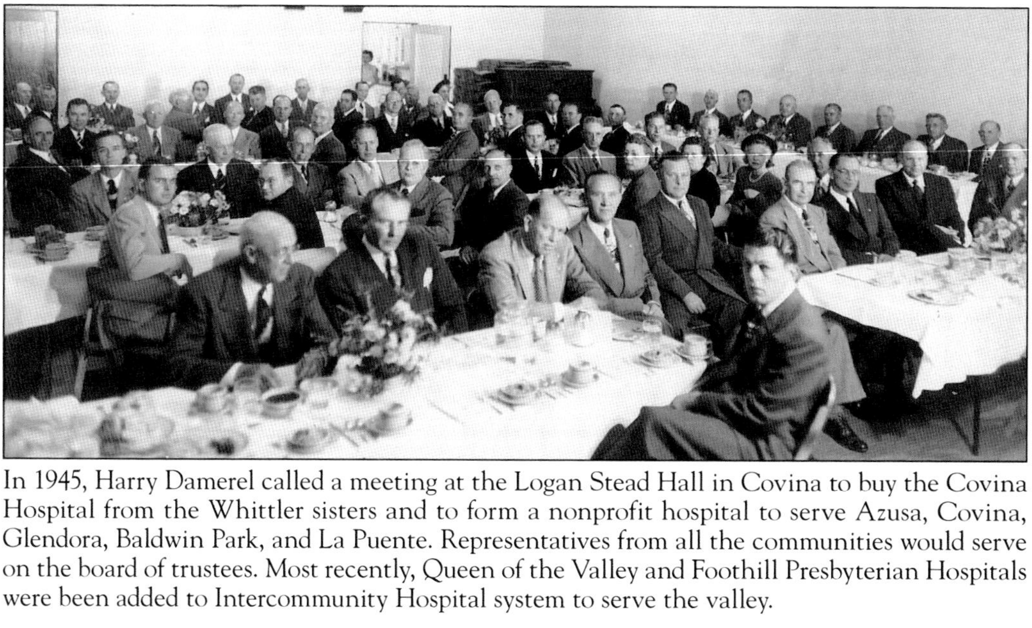

In 1945, Harry Damerel called a meeting at the Logan Stead Hall in Covina to buy the Covina Hospital from the Whittler sisters and to form a nonprofit hospital to serve Azusa, Covina, Glendora, Baldwin Park, and La Puente. Representatives from all the communities would serve on the board of trustees. Most recently, Queen of the Valley and Foothill Presbyterian Hospitals were been added to Intercommunity Hospital system to serve the valley.

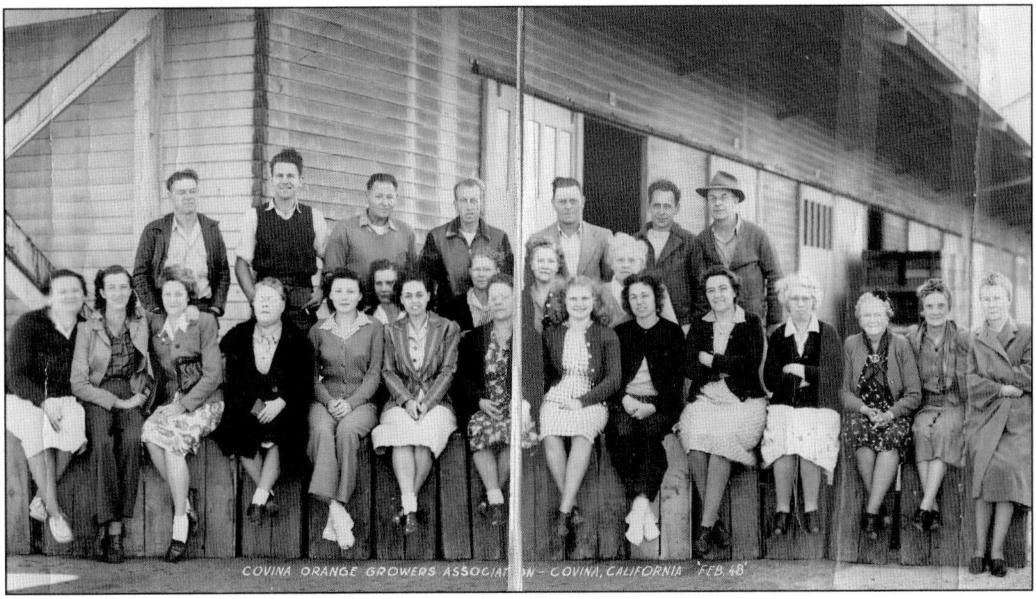

In February 1948, the Covina Orange Growers Association held a reunion at its packing plant for everyone who had worked at the plant over the years. This picture was taken outside. The women are seated on orange crates. Only several of the workers are identified: Betsy Jackson (first row, far left), Mary Coolman (first row, second from left), and Anna Miller (first row, fourth from left).

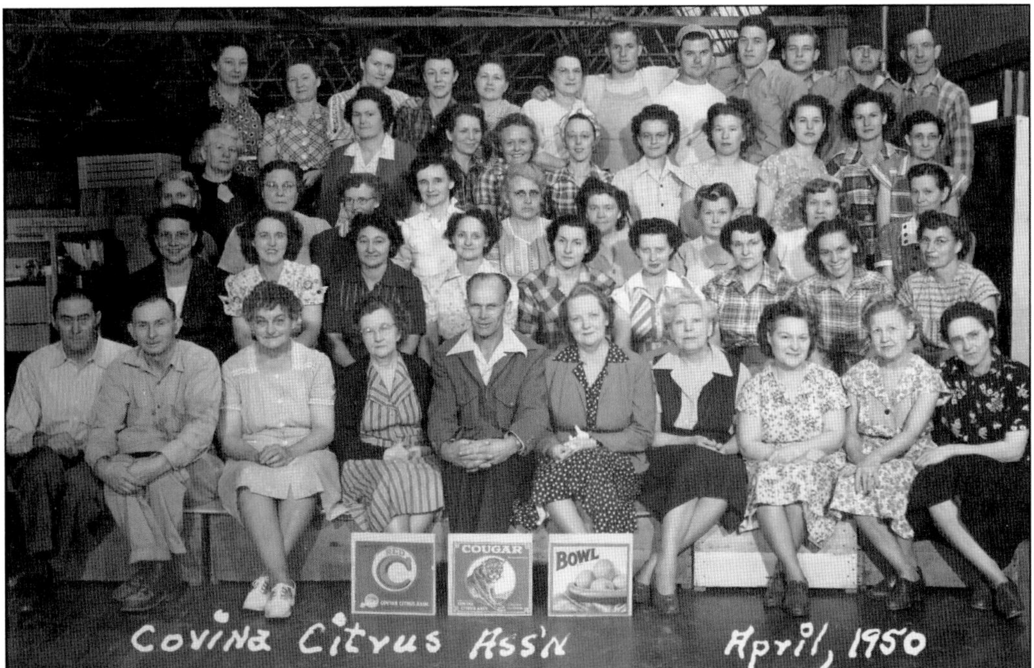

In April 1950, the Covina Citrus Association also sponsored a reunion. Vollie Moore donated this photograph to the Covina Valley Historical Society. For the photograph, they constructed temporary bleachers from planks and packing crates. In front are three packing crates with the Red C, Cougar, and Bowl labels. Sitting sixth from the left in the first row is Charlotte Naftel. On the fifth row, the first and second workers from the left are Mrs. Chafee and Laurena Mae Smith.

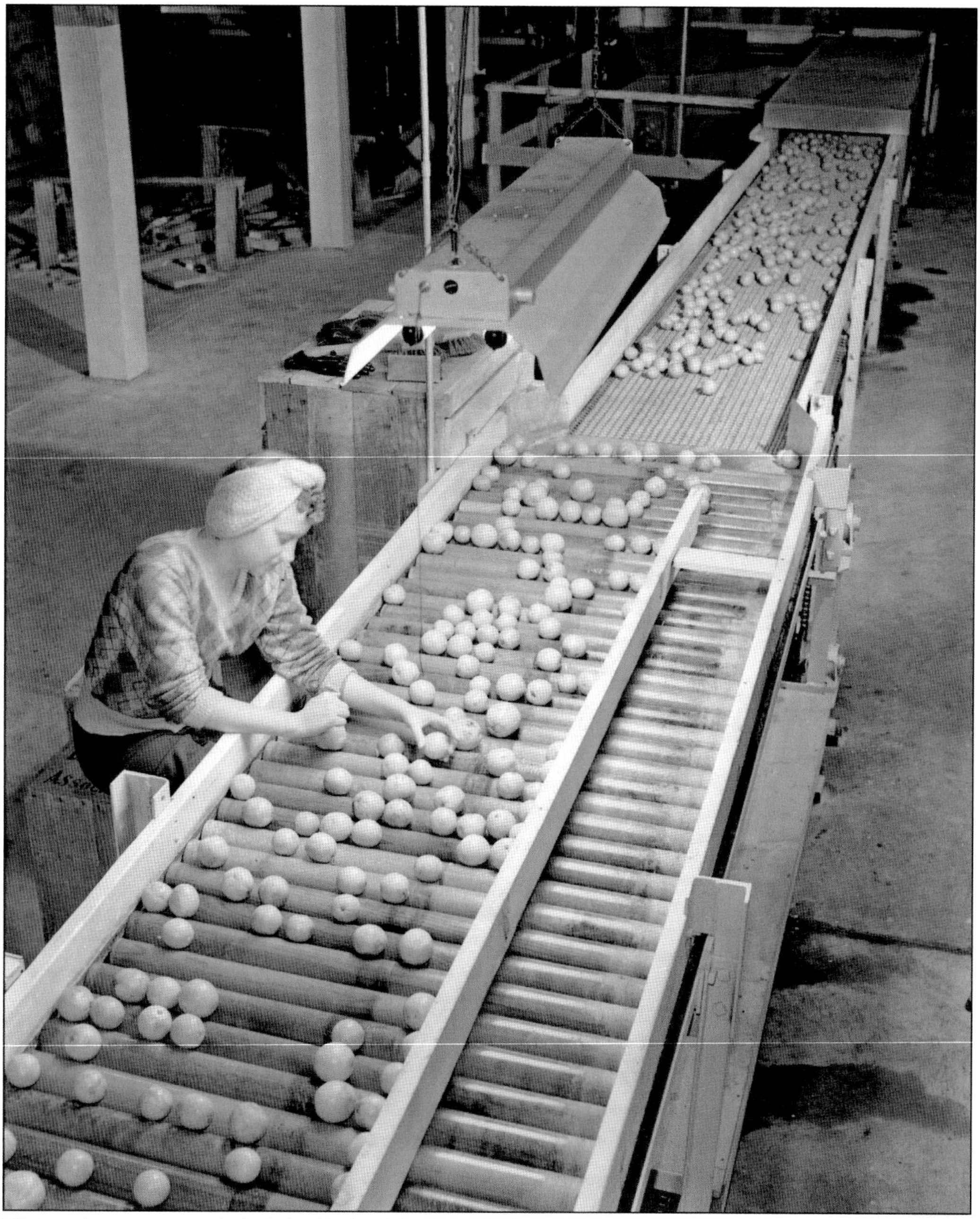

This sad picture entitled, "The End of the Line," shows a lone woman sitting on a field box; she is contemplating the oranges left in a packinghouse that was being dismantled before closing. A tribute to the pioneers who developed the Covina Valley citrus industry from the June 1902 issue of *Resources of Southern California* is appropriate here: "Orange growing is a unique industry, of which Americans should feel proud, for it has been built up in the face of inexperience and of apparently insurmountable obstacles by true American grit, ingenuity and perseverance."

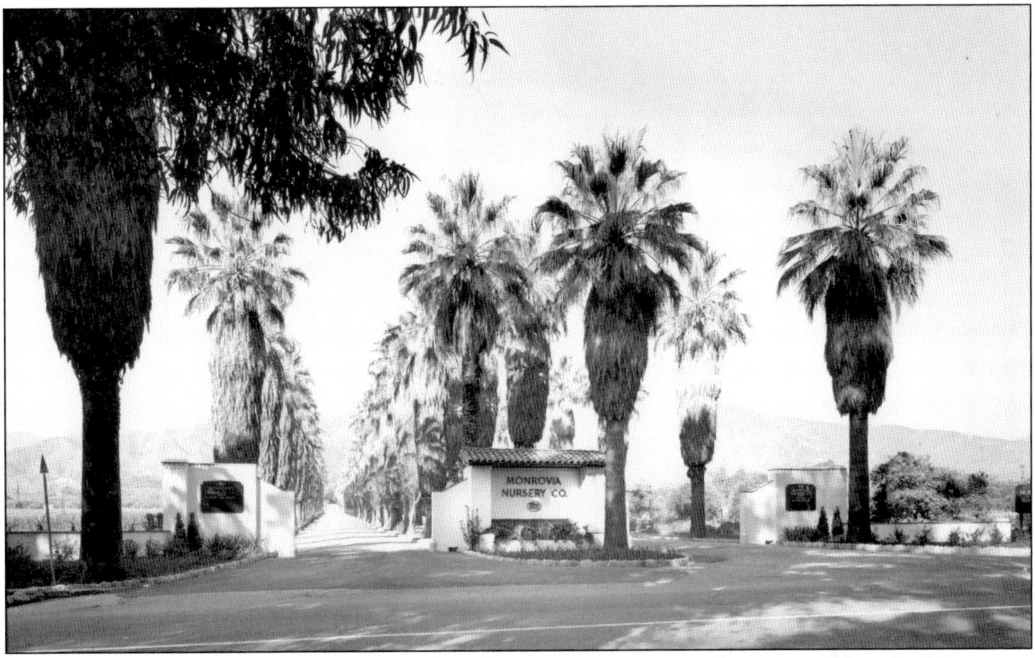

In 1954, the Macneil's Alisal rancho and Vosburg's Cacomites rancho were sold to the Monrovia Nursery Company. Harry Rosedale, founder and owner of the company, had the California Mission–style entrance built to unite the two ranchos. The entrance became a symbol for the company and appeared in sales catalogs. (Courtesy of Jeffrey Cornejo Jr.)

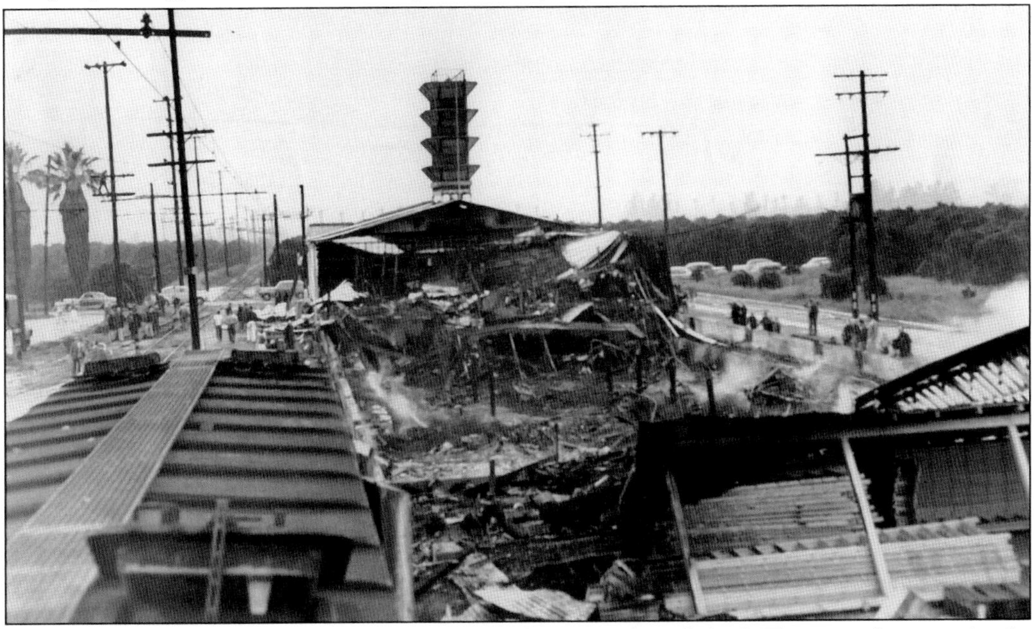

In the 1950s, many valley packinghouses burned or were demolished. This is a photograph of a fire-razed packinghouse east of Citrus Avenue in Covina. In July 1951, the worst fire in Covina's history destroyed the Damerel-Allison juice plant and main offices. The company had been sold the year before to Sunkist Lemon Division for $1 million. They expanded their Corona plant and never rebuilt in Covina.

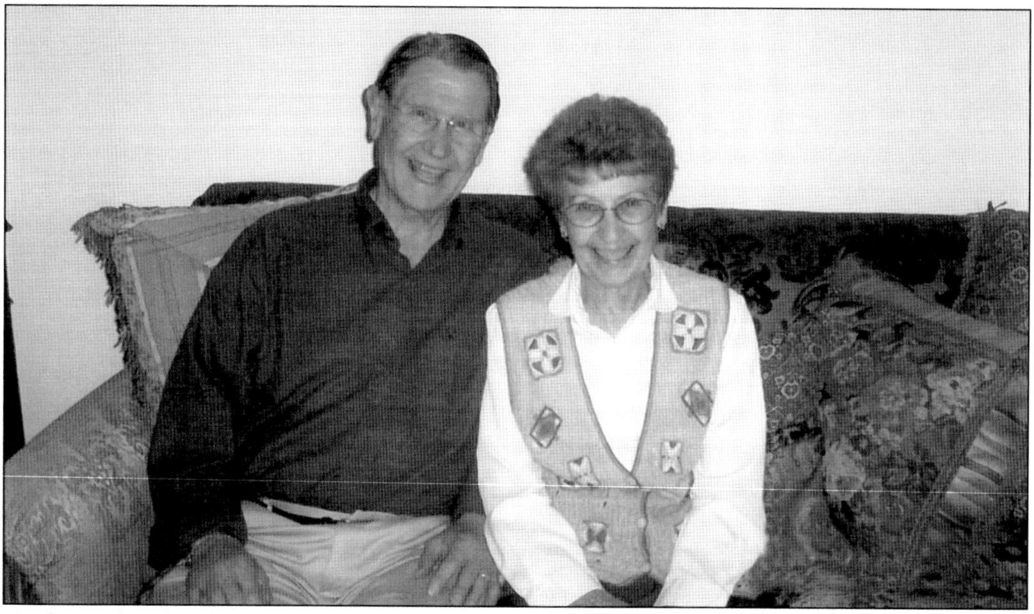

Wesley and Richardene Berry, pictured, became citrus label collectors through Leonard Daniels. Independent packers were their special interest. They joined the California Citrus Label Society. Wesley was vice president of First Federal Savings and Loan and a Host Lions Club member for 49 years. They were also active members of the Covina Christ First Baptist Church.

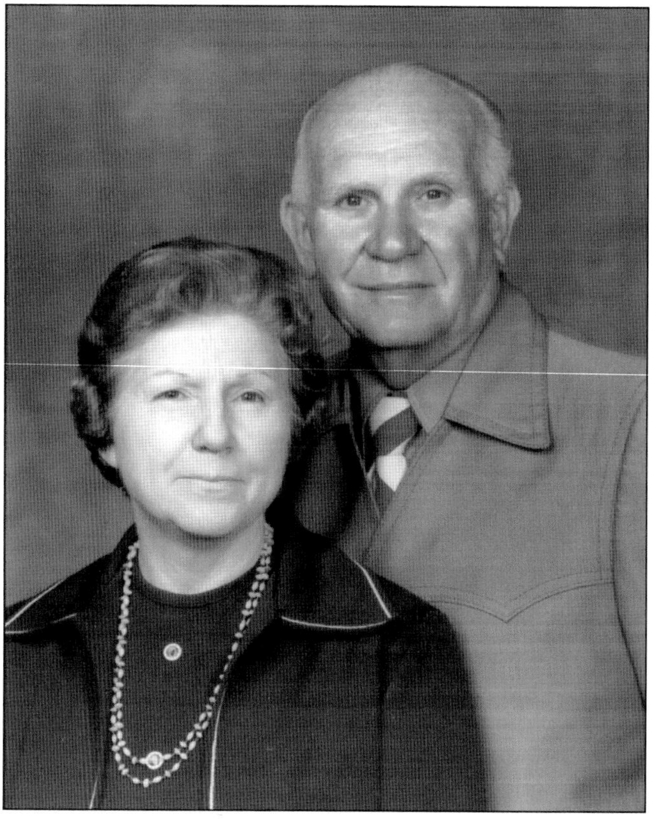

Leonard Daniels Jr., shown with his wife, Aline, had a lifelong interest in citrus crate labels. After attending Chaffey College, Daniels became a citrus rancher, but when the citrus industry began to decline, he went into real estate. He established an office in Charter Oak. When California packinghouses began to close, he traveled throughout the state to collect citrus crate labels that would have otherwise been thrown in the trash.

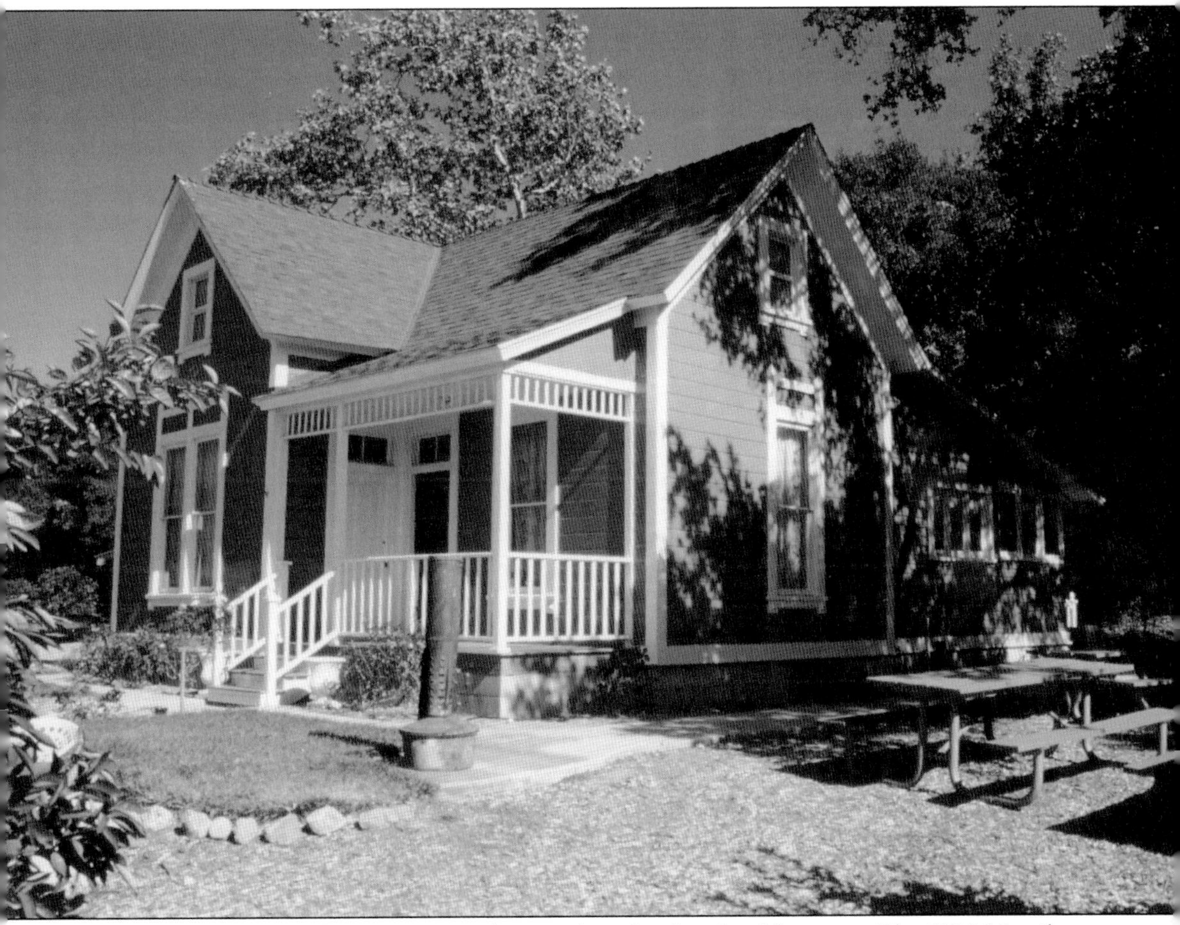

Glendora still has a working citrus ranch open for school and public tours. The 1890 Hamilton House is surrounded by five acres of citrus. The Centennial Heritage Park was created by the Glendora Preservation Foundation. The equipment needed to maintain and operate a citrus ranch is on display. In the garage behind the house, Orton Englehart invented the first action impulse sprinkler. Different designs he developed, including the famous model 60 that was patented in 1935, are on display. Citrus-era memorabilia may also be seen at the Azusa Historical Society Museum, the Covina Firehouse-Jail Museum, the historic photography exhibition at Covina City Hall, and the Glendora Historical Society Museum. Truckloads of oranges are still arriving at the Vita Pakt Products Company located on First and Front Streets in Covina; it is the one remaining working citrus plant in the valley. (Courtesy of Glendora Historical Society.)

www.arcadiapublishing.com

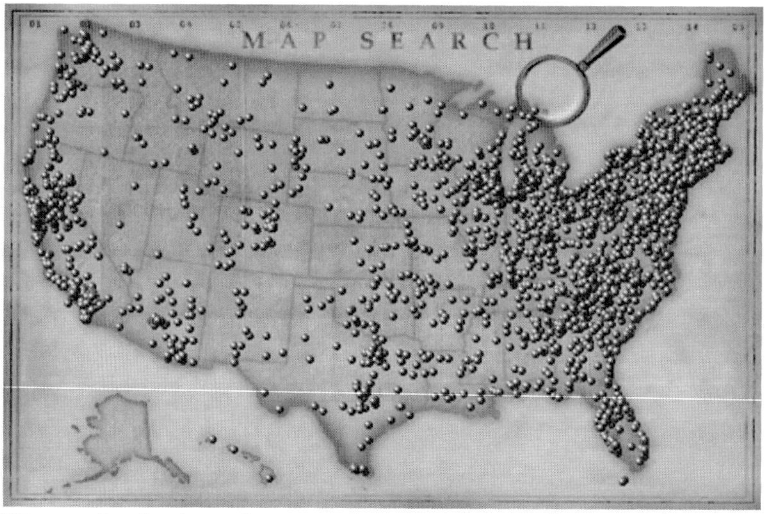

Discover books about the town where you grew up, the cities where your friends and families live, the town where your parents met, or even that retirement spot you've been dreaming about. Our Web site provides history lovers with exclusive deals, advanced notification about new titles, e-mail alerts of author events, and much more.

MADE IN THE USA

Arcadia Publishing, the leading local history publisher in the United States, is committed to making history accessible and meaningful through publishing books that celebrate and preserve the heritage of America's people and places. Consistent with our mission to preserve history on a local level, this book was printed in South Carolina on American-made paper and manufactured entirely in the United States.

This book carries the accredited Forest Stewardship Council (FSC) label and is printed on 100 percent FSC-certified paper. Products carrying the FSC label are independently certified to assure consumers that they come from forests that are managed to meet the social, economic, and ecological needs of present and future generations.

FSC
Mixed Sources
Product group from well-managed forests and other controlled sources

Cert no. SW-COC-001530
www.fsc.org
© 1996 Forest Stewardship Council

Find Your Place in History.